Integrating Sustainability and Autonomous Transportation for Enhanced Urban Mobility

Ergunova Olga Titovna
Peter the Great St. Petersburg Polytechnic University, Russia

Dinesh Elango
University of Arizona, USA

Pawan Kumar
Lovely Professional University, India

Preeti Mehra
Campbell College, Canada & Lovely Professional University, India

Rajesh Verma
Lovely Professional University, India

IGI Global
Scientific Publishing
Publishing Tomorrow's Research Today

Vice President of Editorial — Melissa Wagner
Director of Acquisitions — Mikaela Felty
Director of Book Development — Jocelynn Hessler
Production Manager — Mike Brehm
Cover Design — Jose Rosado

Published in the United States of America by
IGI Global Scientific Publishing
701 East Chocolate Avenue
Hershey, PA, 17033, USA
Tel: 717-533-8845
Fax: 717-533-7115
Website: https://www.igi-global.com E-mail: cust@igi-global.com

Library of Congress Cataloging-in-Publication Data
Names: Ergunova, Olga Titovna, 1980- editor | Elango, Dinesh, 1988- editor | Kumar, Pawan, 1980- editor | Mehra, Preeti, 1981- editor | Verma, Rajesh, 1974- editor
Title: Integrating sustainability and autonomous transportation for enhanced urban mobility / edited by, Ergunova Olga Titovna, Dinesh Elango, Pawan Kumar, Preeti Mehra, Rajesh Verma.
Description: Hershey, PA : IGI Global Scientific Publishing, [2026] | Includes bibliographical references and index. | Summary: "The book aims to address a critical research gap in the literature by focusing on the need to provide an adaptive and sustainable mobility system across different urban eco-systems globally"-- Provided by publisher.
Identifiers: LCCN 2025013363 (print) | LCCN 2025013364 (ebook) | ISBN 9798337308821 hardcover | ISBN 9798337308838 paperback | ISBN 9798337308845 ebook
Subjects: LCSH: Urban transportation--Environmental aspects | Urban transportation--Technological innovations | Sustainable transportation http://id.loc.gov/authorities/subjects/sh2020008871
Classification: LCC HE305 .I4937 2026 (print) | LCC HE305 (ebook) | DDC 388.4--dc23/eng/20250610
LC record available at https://lccn.loc.gov/2025013363
LC ebook record available at https://lccn.loc.gov/2025013364
British Cataloguing in Publication Data
A Cataloguing in Publication record for this book is available from the British Library.

Table of Contents

Detailed Table of Contents

An increment of 1.3 billion in the world's population every 20 years signals to us that in the future, the accumulation of a vast number of people in certain regions and places will lead to land scarcity and carbon emissions issues. This article aims to examine the plans and policies of three countries categorized as developed economies, two countries classified as economies in transition, and three countries considered developing economies. These countries were selected based on data from the World Economic Situation and Prospects 2023. The results of the content analysis enable us to identify the existing and significant disparities between the implementation of plans and policies in the selected countries. More prosperous and developed countries will be more advanced and geared towards electrification in transportation, autonomous and connected vehicles, and artificial intelligence in transportation compared to less developed and less advanced countries.

This research study performs a bibliometric analysis of the evolution of green energy and sustainable mobility using the PRISMA technque. The dataset, spanning from 1999 to 2024, was analyzed using VOSviewer and Bibliometrix software to identify key trends, influential authors, and institutional contributions. Finding of the study demonstrate the expanding academic interest in sustainability, reflected in the increasing number of publications over the years. The research is predominantly led by institutions in North America and Europe, with the United States leading publications, followed by the United Kingdom, China, Canada and Australia. Leading journals such as Sustainability Switzerland, Energies, and Sustainable Cities and Society emphasize the interdisciplinary nature of the field. The study brings out research clusters around key authors like B.K. Sovacool and D. Pamucar. Aligning with UN SDG 12, the study emphasizes the AI-driven innovations in responsible consumption, offering insights for policymakers and managers aiming to integrate AI for sustainable consumer engagement.

The availability of new opportunities and the constant growth of the megalopolis challenge advertisers' sustainable travel based on transport, the volume of which, according to experts, has become one of the main factors affecting greenhouse gas emissions, air pollution, and energy consumption. This chapter aims to cover a crucial area of research that focuses on city mobility concerning achieving and managing sustainability and its impacts on the environment. It covers the trend towards environmentally friendly means of transport, including EVs, hydrogen vehicles, and using green power in city transport. This chapter also addresses ITS concerns regarding traffic management, congestion, and energy use reduction efficiencies. Describing schemes such as smart infrastructure for green transport, energy-effective design, and zero carbon transport policies, this chapter offers perceptive scenarios on how cities may facilitate the transformation to sustainable low-emission transport systems with a vision for the global solution to climate change and improved environmental sustainability.

Urban mobility determines the vector of economic development of territories. And if the positive aspects of it have been studied quite fully, its negative aspects have not been studied sufficiently. A systemic view of the problem of urban mobility determines not only the formulation of tasks, but also determines the methods of their solution. The chapter uses an interdisciplinary approach, including the analysis of normative documents, statistical data, a comparative study of foreign experience. Content analysis methods of interviews with experts in the field of urbanism were used to study ethical aspects. The results of the study revealed that urban mobility contributes to the deepening of social stratification in a number of cities. The analysis shows that effective urban mobility management should be based on the principles of social justice, environmental sustainability, and ethical responsibility. The chapter provides recommendations for the implementation of inclusive transport solutions and emphasizes the importance of integrating ethical principles into an urban policymaking.

The efficiency of urban mobility systems significantly depends on the following factors: what technical and technological solutions are used in the system, how the interaction of all system participants is organized and what is the main motivation of the system users. This means that when analyzing urban mobility systems, it is necessary to use a socio-technical approach. The work is based on cases of integrating sustainable transport solutions of Peter the Great St. Petersburg Polytechnic University (SPbPU) into the urban transport system of St. Petersburg. Following the socio-technical approach, the chapter presents the results of the analysis of the technical and technological subsystem of the corporate transport system with an emphasis on the infrastructure necessary to support various innovative solutions in the transport sector.

The transportation sector is undergoing a profound transformation driven by emerging technologies such as autonomous vehicles, electrification, IoT, artificial intelligence, blockchain, and robotics. This chapter examines how these advancements are reshaping employment, introducing new technology-focused roles while displacing traditional ones. It delves into automation's impact on workforce dynamics, the rise of gig-based employment, and the skillsets required for future opportunities.

Preface

INTRODUCTION

As editors of this volume, *The Future of Urban Mobility: Integrating Sustainability and Autonomous Transportation for Enhanced Urban Mobility*, we are pleased to present a scholarly collection that addresses one of the most pressing challenges of our time: how to enable sustainable, inclusive, and technologically advanced urban transportation in an era marked by rapid urbanization and escalating environmental concerns.

Today, the growth of urban centers across the globe is undeniable. With this expansion, however, comes a surge in the number of vehicles on the roads—bringing with it issues of pollution, traffic congestion, and social inequities in access to transportation. These problems compel us to critically examine and innovate the way we design and operate our urban mobility systems.

This book emerges as a response to that call. Anchored in the themes of sustainability, technology, and equity, it explores comprehensive solutions for future-ready urban transportation. The volume delves into the integration of Electric Vehicles (EVs), Shared Mobility platforms, and Autonomous Transportation systems, offering a multidisciplinary perspective on the evolution of mobility. By highlighting the transformative potential of these emerging technologies, we aim to spark a shift from conventional, resource-heavy transport models to ones that are cleaner, more efficient, and deeply human-centered.

Our vision is one of accessible urban mobility—free from socio-economic or geographical barriers. Whether in megacities or small urban clusters, all citizens deserve mobility options that are safe, sustainable, and inclusive. We emphasize the urgent need for transportation designs and marketing narratives that uphold the values of diversity, equity, and inclusivity. Every stakeholder—be it governments, private enterprises, or civil society—must contribute to building transportation ecosystems that reflect these principles.

Through global case studies and evidence-based research, this book examines the environmental implications of electric and autonomous vehicles, the design of self-sufficient mobility systems, and the ethical considerations shaping urban mobility's future. It offers actionable insights and best practices that can guide urban planners, policymakers, academics, engineers, and sustainability advocates toward more resilient and adaptive transportation infrastructures.

Our primary objective is to bridge an existing research gap—offering a holistic framework that unites environmental consciousness with technological innovation. We hope this book inspires new discussions and collaborative action in the journey toward greener, smarter, and more equitable urban mobility systems worldwide.

ORGANIZATION OF THE BOOK

This volume opens with a critical examination of global transportation trends and the challenges accompanying rapid population growth. Chapter 1 provides a comparative analysis of transportation policies across developed, transitioning, and developing economies, revealing significant disparities in the adoption of electrification, autonomous, and AI-driven transportation technologies. This foundational perspective underscores the uneven pace of progress and the urgent need for inclusive strategies worldwide.

Building on this, Chapter 2 employs a bibliometric study to map the evolution of green energy and sustainable mobility research over the past quarter-century. By highlighting leading authors, institutions, and research clusters predominantly from North America and Europe, the chapter illuminates how interdisciplinary efforts and AI innovations are driving responsible consumption aligned with the United Nations Sustainable Development Goals.

Chapter 3 turns to the environmental dimension, exploring sustainable urban mobility's role in mitigating greenhouse gas emissions and reducing energy consumption. It reviews advancements in electric and hydrogen vehicles alongside smart infrastructure and traffic management systems, painting a comprehensive picture of how cities can transition toward low-emission transport networks that address the global climate challenge.

The social and ethical implications of urban mobility are the focus of Chapter 4, which adopts an interdisciplinary approach to examine how transportation systems can exacerbate social stratification. The chapter calls for integrating social justice, environmental stewardship, and ethical responsibility into urban mobility policymaking, advocating for inclusive solutions that ensure equitable access for all urban residents.

Recognizing that mobility systems are complex socio-technical entities, Chapter 5 applies this framework through case studies from St. Petersburg, Russia. It analyzes how the interplay between technology, infrastructure, and user behavior shapes urban transport efficiency, highlighting the importance of coordinated design and management of innovative mobility solutions.

Chapter 6 investigates the transformative impact of emerging technologies—including autonomous vehicles, electrification, IoT, and AI—on employment within the transportation sector. It critically examines workforce displacement and the rise of new roles, emphasizing the need for reskilling initiatives and collaborative policy frameworks that balance technological progress with social equity and job security.

In Chapter 7, the editors showcase the potential of graph theory combined with AI and big data to optimize multi-modal transportation networks. This chapter presents cutting-edge approaches to route optimization, congestion management, and resource allocation, underscoring how these innovations can create resilient and sustainable urban mobility systems.

Chapter 8 explores Intelligent Transportation Systems (ITS) as a vital tool for addressing urban congestion and its environmental consequences. Through real-world case studies from Singapore and Barcelona, the chapter demonstrates how real-time traffic management and connected vehicle technologies improve traffic flow and reduce pollution, advocating for integrated ITS deployment in sustainable urban planning.

Addressing infrastructure maintenance, Chapter 9 offers a novel framework that leverages deep learning and GIS technologies to automate pothole detection and prioritization. This data-driven approach enhances road safety and optimizes repair resource allocation, representing a forward-looking solution for proactive urban infrastructure management.

CONCLUSION

As editors, we are proud to present this collection of chapters that collectively illuminate the multifaceted challenges and innovative solutions shaping the future of sustainable mobility and digital transformation in transportation systems worldwide. The contributions span critical technological advancements—from wireless charging innovations and autonomous transport policies to digital literacy and the evolving role of digital competencies in higher education. They underscore the interdependence of technology, policy, and social equity in crafting transportation systems that are not only efficient and resilient but also inclusive and environmentally responsible.

The diverse geographic contexts covered—from emerging economies in Africa, Russia, and India to developed urban centers—highlight that while the pathways may differ, the overarching goals remain universal: reducing carbon footprints, bridging inequality gaps, and fostering economic prosperity through smart, data-driven solutions. Importantly, these chapters affirm that technological innovation alone is insufficient without integrated policies, stakeholder collaboration, and attention to local socio-economic realities.

We hope this volume serves as both a comprehensive reference and a catalyst for further interdisciplinary research, policy dialogue, and practical implementation in the global quest for sustainable, equitable, and future-ready mobility systems. The road ahead demands innovation coupled with inclusivity, and it is our sincere belief that the insights compiled here provide meaningful guidance for researchers, practitioners, and policymakers dedicated to this imperative.

Chapter 1
The Rising Trends and Challenges in World Transportation

Rashid Ating
https://orcid.org/0000-0001-6716-8500
Department of Primary Care Medicine, Faculty of Medicine, Universiti Malaya, Malaysia

April Kim T. Lalisan
https://orcid.org/0000-0002-5942-7637
University of Perpetual Help System Dalta, Philippines

ABSTRACT

An increment of 1.3 billion in the world's population every 20 years signals to us that in the future, the accumulation of a vast number of people in certain regions and places will lead to land scarcity and carbon emissions issues. This article aims to examine the plans and policies of three countries categorized as developed economies, two countries classified as economies in transition, and three countries considered developing economies. These countries were selected based on data from the World Economic Situation and Prospects 2023. The results of the content analysis enable us to identify the existing and significant disparities between the implementation of plans and policies in the selected countries. More prosperous and developed countries will be more advanced and geared towards electrification in transportation, autonomous and connected vehicles, and artificial intelligence in transportation compared to less developed and less advanced countries.

DOI: 10.4018/979-8-3373-0882-1.ch001

1: INTRODUCTION

Transport connectivity is a crucial prerequisite for tourism (Kim, 2019, as cited in UNWTO, n.d.). Based on the United Nations World Tourism Organization's definition of tourism, tourism necessarily entails the movement of people. With this, one can recognize the significant role of transportation in the industry. Without transportation, mobility will not take place. Mobility is essential for an individual to reach a tourist destination. According to Rodrigue (2020), Tourism, as an economic activity, relies on transportation to bring tourists to destinations, and transportation itself can be part of the tourist experience.

Over the decades, tourism has experienced continued growth and deepening diversification, becoming one of the fastest-growing economic sectors globally (UNWTO, n.d.). Tourism has long been recognized as a significant contributor to the global economy. According to the UNWTO (n.d.), international tourist arrivals (overnight visitors) grew 4% in January-March 2019 compared to the previous year, below the 6% average growth of the past two years. However, in 2020, the COVID-19 pandemic profoundly negatively impacted the tourism industry, with international tourism figures dipping significantly.

After the unprecedented drop in 2020 and 2021, international tourism is expected to continue its gradual recovery in 2022 (UNWTO, 2022). With the expected recovery, the availability of transportation services will be crucial in achieving its recovery as transportation is the means that provides access to tourist destinations.

Innovations in long-haul air transportation made international tourism possible soon after, which allowed for tourism in previously unreachable destinations (Chew, 1987; as cited in Van Truong and Shimizu, 2017). Rodrigue (2020) also supports this statement, mentioning that the expansion of international tourism significantly impacts the discipline of transport geography, as it links traffic generation, interactions at different scales (from local to global), and related transportation modes and terminals.

The tourism industry is dependent on the transportation industry. Although technological innovations have led to the rise of the tourism industry, it is not without consequences. The tourism sector is highly vulnerable to climate change and, at the same time, contributes to the emission of greenhouse gases (GHG), which cause global warming. Therefore, accelerating climate action in the tourism sector is of utmost importance for its resilience (UNWTO, n.d.). With COVID-19 hampering the growth of the tourism industry, one might think it poses the greatest threat. However, climate change, not the pandemic, is the biggest threat to tourism. UNWTO's Secretary-General Zurab Pololikashvili has warned that the "climate emergency is a bigger threat than Covid" (UNWTO, 2021b).

The United Nations (1992) defines climate change as a change of climate attributed directly or indirectly to human activity that alters the composition of the global atmosphere in addition to natural climate variability observed over comparable periods. As greenhouse gas emissions blanket the Earth, they trap the sun's heat, leading to global warming and climate change (UN, n.d.). According to the United States Environmental Protection Agency (USEPA, 2019), the largest

source of greenhouse gas emissions from human activities in the United States is burning fossil fuels for electricity, heat, and transportation.

Transportation is essential in all industries. However, transportation seems a double-edged sword that brings good and evil. Since the tourism industry relies heavily on transportation, it is essential to examine the trends in transportation and the challenges this sector faces. This study aims to discover the rising trends and challenges in world transportation.

2: STATEMENT OF THE PROBLEM

1. What are the transportation trends in the world?
2. What are the transportation challenges in the world?

   What are the plans, and policies of the chosen countries in terms of

   a. Electrification in Transportation,
   b. Autonomous and Connected Vehicles, and
   c. Artificial Intelligence in Transportation?
3. What are the measures taken by the chosen countries in terms of climate change mitigation and adaptation?

3: TRANSPORTATION TRENDS

Affordable air travel, increased connectivity, new technological advances, new business models, and excellent visa facilitation worldwide have fostered the continuous growth of international and domestic tourism in the past decades (UNWTO, 2019). Tourism is seen as a significant contributor to the Sustainable Development Goals 2013. However, increasing traffic due to world tourism growth can adversely affect congestion, safety and security problems, pollution, etc. (Pellegrino, 2021).

Transport is important for tourism because it a) facilitates the movement of tourists between their place of origin and their destinations, and b) acts as the means of movement within a destination, thus allowing for wider dispersal of visitor movement and, as a result, maximum exposure of visitor flows to areas perhaps not otherwise possible (Page, 2009, as cited in Lohmann, and Duval, 2011). Transportation is a fundamental driver of the tourism industry: it is a precondition for travel since it facilitates mobility and the movement of tourists from their place of origin (i.e., their home area) to their destination and back (Page and Ge, 2009). Transportation is the primary means to carry passengers, the tourists, to the actual site where tourism services are performed (Mammadov, 2012).

3a: Electrification

Transportation is a significant contributor to greenhouse gas emissions. According to Sustainable Travel International, tourism is responsible for roughly 8% of the world's carbon emissions (n.d.). With this, the industry must develop steps to reduce its emissions. One of the trends to address this issue is the electrification of transportation. Electrification of transportation, primarily through the adoption of personal electric vehicles (EVs), is increasing worldwide (O'Neill-Carrillo, Lave, and Haines, 2021). As of 2020, EVs accounted for just 6 per cent of global automotive sales, but that is projected to grow to 13 per cent by 2025 and 22 per cent by 2030 (Marr, 2022).

Pursuing electrification in transportation may lead to a reduction in greenhouse gas emissions. By reducing its carbon footprint in the transport and accommodation sector, tourism can benefit from low carbon growth and help tackle one of the most pressing challenges of our time (UNWTO, 2017). One of the solutions advanced by the motor vehicle industry globally to counter the instability of prices has been its inroads in the electrification of cars with a wide variety of electric and hybrid vehicles now available on the market (Fjelstul, 2014).

Investment in green infrastructure (more efficient transport, and reduced air pollution) should result in more innovative and greener cities for residents and tourists (UNWTO, 2017). Many countries are slowly transitioning to the adoption of electrification of transportation systems, and one example is Norway. Norway is already a global leader in adopting electric vehicles, spurred in considerable measure by the hydropower that provides 98 per cent of the country's electricity (Hockenos, 2018).

The year 2019 was a record-breaking year for electric vehicle sales, but the pandemic and economic downturn slowed growth in the United States and around the world (Colonel, 2020). In the article by Forbes, Marr (2022) mentioned that over time, stricter national emissions targets, more significant urban populations, improvements in charging infrastructure, and the declining cost of the lithium-ion batteries that power EVs (already down 80 per cent since 2010) will combine to encourage mass adoption of EVs.

3b: Electrification in Air Transportation

Air transport is considered tourism's main contributor to climate change (UNWTO, 2018). The share of air travel has increased from 46% in 2000 to 59% in 2019, while land transport has decreased from 49% to 35% in the same period (UNWTO, 2021a). Air transportation is the primary mode of transportation in international tourism; steps must be taken to minimize its negative impact on the environment. However, current carbon dioxide (CO2) emissions from commercial aviation are estimated to be 186 million metric tons (MMmt) CO2e, and CO2 emissions are projected to grow to 209 MMmt CO2e in 2050 (Environmental Investigation Agency, 2020, as cited in Schwab et al, 2021).

On the website of Sustainable Travel International, planes and cars generate the most CO2 per passenger mile, with tour buses, ferries, and trains coming well behind. In recent years, the number of people travelling internationally skyrocketed as airfare became more affordable. Similarly, between 2005 and 2016, transport-related tourism emissions increased by more than 60% (n.d.).

Electrified aviation covers a wide range of aircraft types and varies in the extent of and approach to electrification. Classes of electrification include what we call here more electric, hybrid electric, and fully electric (Schwab et al, 2021). Electrification of air transportation is not impossible. According to Rolls Royce, the organization is championing electrification – developing, testing, and innovating technologies and systems together with partners to create tomorrow's electric propulsion and energy systems (n.d.). The electrification of air transportation is the third era of aviation. Aviation's third era is enabled by advances in new architectures, advanced engine thermodynamic efficiencies, electric and hybrid-electric propulsion, digitization, artificial intelligence, materials, and manufacturing (Jarayaman, 2019).

According to the consultancy Roland Berger, the number of electric aircraft in development increased by roughly 50% over the past year to 170. The number could swell to 200 by 2019 (Riley, 2019). Moreover, in the CNN Business article, Riley emphasized the following developments:

> *Two significant factors drive increased investment: The global aviation industry produces up to 3% of all carbon dioxide (CO2) emissions, a share projected to rise sharply in coming years. It spends roughly $180 billion a year on jet fuel.*
>
> *"Increased efficiency has been the name of the game when selling aircraft for decades," said Nikhil Sachdeva, a senior consultant at Roland Berger. "Electric is the next phase."*
>
> *United Technologies (UTX) announced plans to merge with defense contractor Raytheon (RTN) just before the Paris Auto Show.*

Moreover, in the article published on the website of Business Standard by MacEacheran (2019), the following developments were expressed:

Rolls-Royce is planning to trial its Accel plane next year (with a top speed of 480 kph [298 mph]), and Israel's venture capital-backed Eviation has predicted delivery of its e-plane to its first customer, a US regional airline, by 2022. With E-Fan X, the next step in Airbus's long-anticipated electrification journey, the European aerospace giant is betting on hybrid.

Electrification of the aviation industry will require more than just the development of electric-powered aircraft. If commercial airline carriers adopt electric aircraft within their fleets, charging events at airports would increase the airports' electricity demands (Hou, Bose, and Haran, 2020). The transition away from jet fuel-powered conventional airplanes towards battery-powered electrified aircraft will impose extra charging requirements on airports (Hou, et al., 2021). A research paper by Mitici, Pereira, and Oliviero (2022) stated that from an operational point of view, however, replacing conventional aircraft with e-AC for short-range flights poses several challenges:

First, how many electric aircraft should an airline acquire to satisfy a given flight demand? Second, the aircraft are expected to be charged at the airport, using charging stations and/or use fully charged spare batteries. Also here, it is of interest to understand how many charging stations an airport would need to sustain electric flights, as well as how many spare batteries are needed. Third, it is of interest to understand how the e-AC should be scheduled for flight and battery recharging.

With the developments, it is crucial to look at how to address the presented challenges in the electrification of air transportation. Innovative solutions together with policies and legislation will help all of these materialize.

3c: Electrification in Land Transportation

Finding a way to reduce greenhouse gas emissions dramatically is top of the agenda for most countries, and many governments see electric cars as part of the solution (Bateman, 2021). Electrification is widely considered an attractive solution for reducing the oil dependency and environmental impact of road transportation (Zhang and Fujimori, 2020). Significant greenhouse gas reductions are possible with a fully decarbonized grid and electric land transport (Nadolny et al., 2022). Renewable electrification of land transport via electric vehicles (EV) would reduce greenhouse emissions by 15–25% globally (International Energy Agency, 2018, as cited in Nadolny et.al., 2022). Road transport electrification will form a substantial part of international strategies to mitigate greenhouse gas and air pollutant emissions (Yulo Guo, 2021).

Looking into transitioning to the use of EVs is beneficial as it helps mitigate climate change. According to Zhang and Fujimori (2020), transport electrification alone would not contribute to climate change mitigation; it is interesting to note that switching to electrified road transport under the sustainable shared socioeconomic pathways permitted an optimistic outlook for a low-carbon transition, even in the absence of a decarbonized power sector. With this, it is indeed promising to shift to using Evs.

With the potential of Evs in helping mitigate climate change, one might ask why countries haven't adapted to their use. A UK survey found the most common reason for not buying one was a lack of fast charging points (37%), followed by concerns about range (35%) and cost (33%) (Broom, 2021). In the same article by Broom, it was mentioned by Richter (2021) that although two-thirds of Americans surveyed by Pew Research agreed electric cars are good for the environment, a similar proportion thought they were expensive. A third were concerned about their reliability (2021). However, although gradual, the shift was seen to rise as presented by Electric vehicles accounted for 4.2% of light vehicle sales in 2020, almost double 2019's 2.5%.

Figure 1. Electric mobility in different countries

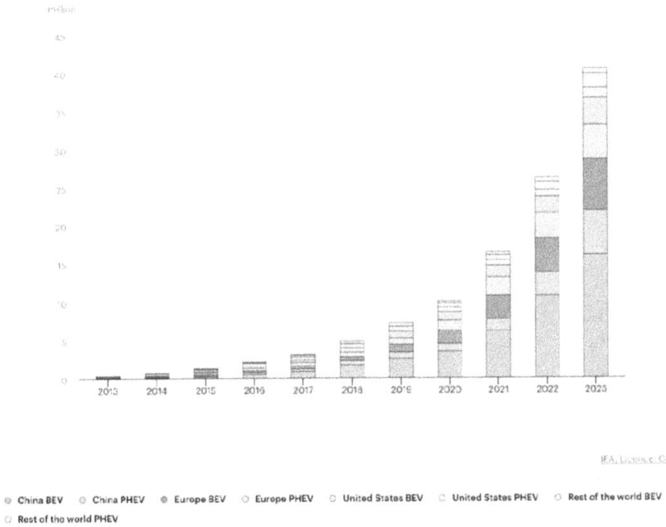

Figure 1 shows the distribution of global electric car stock, from 2013 to 2023. Worth noting that the demand for this electric car is keeps increasing, and the same trend is projected is happen in the future with the advancement of technology, particularly with the introduction of Artificial Intelligence (AI) in the automobile industry. The same chart also offers some insight into the current development of Battery Electric Vehicle (BEV) and Plug-in Hybrid Electric Vehicle (PHEV) in China (as a newcomer in this industry, as reported there is an enormous number of companies originally from other industries jumping into this industry such as Zeekr, Xiomi, XPeng, Chery and others due to the government support, aligned with the China President narrative, during his inspection trip to Helionjiang province in September 2023. While visiting this province, Xi quotes the term "New Quality Productive Force", with electric vehicles as one of the three focus industries) against the established companies either from Europe or the USA. Nonetheless even with generous support from the Central Communist Party of China, to promote the industry, especially in electric vehicles as one of the key areas in New Quality Productive Forces, the reality proves too harsh. In 2019, according to the report from Bloomberg, there were 500 Chinese electrics electric car manufacturing companies were established, and after going through fierce competition, only 100 electric car manufacturers remained in 2023, while the rest were out of competition or declared bankrupt.

The same figure shows that even with the remaining 100 or fewer electric car manufacturers, we can say that Chinese BEVs and PHEVs are dominating the world market now. Not an exaggeration to say that the volume of Chinese car electric vehicles produced surpasses both the EU's and the USA's combined total value produced. This Chinese electric car manufacturer is not only successful in the domestic market but in the international market also. One of the prime examples is BYD (Build Your Dream), a Chinese electric car manufacturer, that has expanded its market, not only in China but in a few Southeast Asia (SEA) countries such as Malaysia, Thailand, Indonesia and the rest.

3d: Electrification in Water Transportation

Water transportation is one of the major modes of transportation. In the tourism industry, water transportation is used to transport goods and people. Moreover, cruise ships which are part of water transportation systems serve as an attraction for tourists. It has been found that 23% of the total vessels in the last ten years have been particularly designed for tourism (Euro News, 2020, as cited in Anwar et al., 2020).

The electrification of vessels/ferries for green transformation requires onboard electrical energy storage as well as an energy supply network in the port area (Anwar et al., 2020).

3e: Electrification of Vehicles Used in the Tourism Industry

Mobility plays a crucial role in destination sustainability (Scuttari, and Isetti, 2019). In recent years the potential positive effects of the electrification of mobility – such as the reduction of oil dependency, CO2 emissions, air pollution and noise – have been widely highlighted by policymakers, industries, and the scientific community (Grauers et. al., 2013, Larminie. & Lowry, 2012, as cited in Scuttari, and Isetti, 2019) as well as several supranational organizations (see e.g., European Green Vehicles Initiative – https://egvi.eu/, European Alternative Fuels Observatory – https://www.eafo.eu/, European Environment Agency – https://www.eea.europa.eu/, International Energy Agency – https://www.iea.org/). While its electrification would on one hand mitigate some environmental and acoustic side-effects, on the other, it poses several technological and governance-related challenges (Scuttari, and Isetti, 2019).

An example of a government initiative is the Bali Government, which has implemented the electric vehicle (EV) policy as a high priority considering its attractiveness for emission and air pollution reduction to maintain the sustainability of Bali's nature and tourism sector (Rizki, 2021).

3f: Autonomous and Connected Vehicles

Connectivity and automation are two separate forms of technology that are often mentioned in the same breath - connected and autonomous vehicles (or CAVs) (Brake, n.d.). Brake is a road safety charity in the United Kingdom defined as a connected vehicle that can share information (or data) with other sources both inside and outside of the vehicle (n.d.). Moreover, it also defines automated technologies as those which undertake parts of the driving task, or in the case of full autonomy, the whole driving task (hence the use of the term, 'driverless car').

Brake on their website, stated that connected and autonomous vehicles have the potential to eliminate human error from driving, a factor in nearly 9 in 10 crashes on UK roads, and assist traffic management, meaning that congestion and pollution are reduced. Moreover, fully driverless vehicles also offer potential benefits for those with additional mobility needs, such as the elderly or disabled people (n.d.).

It is said that the use of advanced connectivity can help car manufacturers increase fuel efficiency by optimizing the route and drive cycle of the vehicle over the whole projected journey (Ricardo, n.d.).

Burges and Salmon, an independent United Kingdom law firm identified the levels of automation of CAVs as: No Automation (Level 0), Driver Assistance (Level 1), Partial Automation (Level 2), Conditional Automation (Level 3), High

Automation (Level 4), and Full Automation (Level 5) (n.d.). Burges and Salmon defined the levels as follows:

No Automation (Level 0)

The human driver controls all aspects of driving all of the time.
Driver Assistance (Level 1)
Human driver is assisted with either steering or acceleration/deceleration by the driver assistance system with the expectation that the human driver will perform all remaining functions.

Partial Automation (Level 2)

Driver assistance system undertakes steering and acceleration/deceleration using information about the driving environment with the expectation that the human driver will perform all other driving tasks.

Conditional Automation (Level 3)

An automated driving system undertakes all aspects of the dynamic driving task with the expectation that the human driver will respond appropriately to a request to intervene.

High Automation (Level 4)

An automated driving system undertakes all aspects of the dynamic driving task, even if a human driver does not respond appropriately to a request to intervene.

Full Automation (Level 5)

Automated driving system undertakes all aspects of the dynamic driving tasks in all roadway and environmental conditions.

3g: Autonomous and Connected Vehicles and Tourism

Connected and autonomous vehicles (CAVs) have the potential to disrupt all industries tied to transport, including tourism (Chone and Hopkins, 2018). The University of Surrey and the University of Oxford have examined how Autonomous Vehicles (AVs) may have a substantial impact on the future of urban tourism. According to these two academic institutions, potential benefits include reduced traffic congestion and emissions, improved foreign car hire processes, reduced parking requirements, and cheaper taxi fares (2018).

Moreover, COVID-19 has significantly increased the use of personal vehicles in travel, adoption of self-driving autonomous vehicles can radically transform the travel industry (Ribeiro, Gursoy, and Chi, 2022).

3h: Artificial Intelligence

Beyond making lives easier, Artificial Intelligence can help to make all transport modes safer, cleaner, smarter, and more efficient (European Union, 2019). Artificial Intelligence (AI) can be defined as a technology that powers machines with human intelligence (Forbes.com, 2019). In an article published on the website of Forbes, Joshi (2019) enumerated the use of AI in transportation as follows:

Self-driving vehicles

Autonomous taxis have already started operating in Tokyo. Also, US logistics are embracing autonomous trucks to reap numerous benefits from it.

Traffic management

AI can not only reduce unwanted traffic congestion but also improve road safety and reduce wait times.

Delay predictions

Leveraging data lake technique and computer vision, transportation companies can offer exceptional service to passengers in cutting down passengers' wait times and enhancing their journey experience.

Drone taxis

Pilot-less helicopters present a unique and sophisticated solution to combat carbon emissions, eliminate traffic congestion, and reduce the need for expensive infrastructure planning and building.

The European Union in their report entitled Artificial intelligence in transport: Current and future developments, opportunities, and challenges (2019) presented the pros and cons of AI in transportation:

AI brings great benefits to road transport but also poses serious challenges, especially in mixed-use environments. AI has the potential to make traffic more efficient, ease traffic congestion, free driver's time, make parking easier, and encourage car- and ridesharing.

As AI helps to keep road traffic flowing, it can also reduce fuel consumption caused by vehicles idling when stationary and improve air quality and urban planning. However, lower transport costs and freeing the driver from driving tasks could also lead to more people choosing a car as a transport mode (instead of public transport), and subsequently, increase congestion and air pollution.

In addition, cybersecurity and data privacy are also of particular importance in the development of AI in automated vehicles. Namely, AI-based automated vehicles require access to a lot of data that is often sensitive or protected. If third parties manage to access automated vehicle data without control, the safety of the vehicle, its occupants, and other road users are endangered.

Systems based on AI used in autonomous transportation could significantly improve road safety, since human error (such as speeding, distraction and drunk driving) is involved in more than 90% of accidents on EU roads, in which more than 25,000 people lost their lives in 2017. In addition, in China, they are several EVs car manufacturers that cut corners and this lead to the EVs car explode due to the sub-par quality of materials used for manufacturing.

However, AI also creates new risky situations, as accidents with automated vehicles have demonstrated. In an interim period, when vehicles are increasingly automated but not yet completely autonomous, drivers might be distracted and pay less attention to the road. When a situation arises where the human needs to intervene, a distracted driver can be slow to react. It is simply difficult for humans to maintain effective visual attention during a longer period of automated driving.

At present, therefore, more progress is needed to ensure that fully automated vehicles can safely interact with other road users, perform well under all weather and road conditions, correctly recognize obstacles and understand the environment.

AI also raises various ethical issues. When faced with life-versus-life situations, the question as to how an AI algorithm in a fully automated vehicle should decide how the vehicle reacts divide’s opinion. In 2016, a Mercedes-Benz executive said: 'If you know you can save at least one person, at least save that one. Save the one in the car'. The respondents to a survey published in Science the same year, on the other hand, wanted others to drive vehicles that would sacrifice their passengers for the greater good whilst simultaneously preferring to drive vehicles that protect their passengers (especially family members) at all costs themselves. Where only one passenger would be killed to save 10 pedestrians, most respondents preferred the vehicle to save the pedestrians.

The use of AI also raises the question as to who should make such a decision. Should AI algorithms automate ethical decision-making independently, or should all vehicles have the same ethical settings, or could people who buy a fully automated vehicle could determine such settings themselves? In case an accident arises, liability is another challenge that needs to be addressed.

A clear boundary of liability must be defined for the different levels of automation so that it is possible to identify who is responsible for the accident. This might require changes in legislation, traffic rules, and insurance policies. The current rules assume that when the vehicle is used on the road, there is a human driver on board, whereas automation technology is intended to partially or fully replace the driver, thus shifting the responsibility.

Regarding employment, it is anticipated that AI will contribute to the creation of new jobs, the disappearance of others, and the modification of most. AI can make workers' lives easier, by helping workers in the car manufacturing industry for instance with repetitive, difficult, and dangerous tasks.

AI can also help to manage large amounts of data and help to plan, design, and control road
transport networks, for example. AI can also enable more people (for instance persons with
disabilities) to participate in the labor market or help to cut labor costs (as an example, long driving hours and stopping for a break will no longer be a concern with fully automated fleets). At the same time, some jobs, such as bus, taxi, and truck drivers, might no longer be needed in a further future when fully autonomous vehicles are available. These drivers will need reskilling to find a different job.

The answers that public and private stakeholders provide on these issues will also determine public acceptance and the consequent market penetration of AI technologies in road transport. A recent Eurobarometer survey on autonomous systems showed that respondents are more comfortable with autonomous vehicles transporting goods than travelling in such a vehicle themselves. The study also indicated that younger respondents and respondents with a higher level of education felt more comfortable about autonomous vehicles.

AI is not new to the aviation industry, which, according to a 2018 International Air Transport Association (IATA) report, has already been used in various parts of the business and across the value chain for decades. However, we are now entering a new era in which AI capabilities are reaching heights that will have a major impact on how the aviation business is conducted (European Union, 2019).

3i: Transportation Challenges

Interregional travel is almost always done by air. In 2016, 96.3% of all interregional arrivals were by air. In 2030, it is expected that 96.5% of interregional arrivals will be by air. Intraregional travel (see figure 2.5) is also mostly done by air in all regions, except Europe, which is the region where arrivals by rail are growing the most (from 26 million in 2016 to 95 million in 2030) (UNWTO, 2019). In this literature, the greatest threat to humanity will be included: climate change.

3j: Climate Change

Climate change is at the top of UN priorities and has become a challenge for many organizations and for dealing with climate change mitigation and adaptation. There is also a common understanding of the need to increase the capacity of countries to deal with this challenge (United Nations Economic Commission for Europe, n.d.).

According to UNECE (n.d.), at several international conferences, transport ministers have addressed the need for CO2 abatement and improved fuel efficiency in the transport sector, mainly through:

Innovative vehicle technologies, advanced engine management systems and efficient vehicle powertrains.

The use of sustainable biofuels, not only of the first generation (vegetable oil, biodiesel, bio-alcohols and biogas from sugar plants, crops or animal fats etc.), but also of the second (biofuels from biomass, non-food crops including wood) and third generations (biodegradable fuels from algae);

An improved transport infrastructure together with Intelligent Transport Systems (ITS) to avoid traffic congestion and to foster the use of intermodal transport (road, rail and waterways);

Consumer information (campaigns for eco-driving*, use of public transport and modal transport etc.);

Legal instruments (such as tax incentives for low-carbon products and processes, taxation of CO2-intensive products and processes, etc.).

4: METHODOLOGY

This study will be using a descriptive research design utilizing secondary data. The researcher will use purposive sampling in choosing the countries to be compared. Three countries per continent (if applicable), except Antarctica, will be chosen. Each country chosen from each continent will represent developed economies, economies in transition, and developing economies. The reference for each category will be the list provided in the World Economic Situation and Prospects 2020 by the United Nations New York.

Once a country is chosen, the researcher will then look at the availability or existence of plans programs, and policies in terms of variables such as the presence and/or adaptation of Electrification, Autonomous and Connected Vehicles, and Artificial Intelligence. The research will also investigate plans or programs, and policies in terms of climate change mitigation and adaptation. The researcher will also investigate climate change mitigation and adaptation measures taken. Points will also be assigned should these be available. The purpose of assigning points to each country is to easily compare the performance of each country in terms of trend adaptation and challenge mitigation.

The following table contains the countries chosen to represent each category:

Table 1. Chosen Countries based on the World Economic Situation and Prospects 2020

CONTINENT	DEVELOPED ECONOMIES	ECONOMIES IN TRANSITION	DEVELOPING ECONOMIES
Africa	N/A	N/A	Egypt
Asia	Japan	Armenia	Philippines
Australia/ Oceania	New Zealand	N/A	N/A
Europe	Norway	Albania	N/A
North America	United States	N/A	N/A
South America	N/A	N/A	Argentina

Source: World Economic Situation and Prospects, (2020).

5: RESULTS AND DISCUSSION

Based on the secondary data gathered online, the following are the results:

Table 2. Electrification in Transportation

COUNTRY		ELECTRIFICATION	
		PLANS/ PROGRAMS	POLICIES/LEGISLATION/ LAW
DEVELOPED ECONOMIES	Japan	✓	✓
	New Zealand	✓	✓
	Norway	✓	✓
	United States	✓	✓
ECONOMIES IN TRANSITION	Armenia	✓	✓
	Albania	✓	✓
DEVELOPING ECONOMIES	Egypt	✓	✓
	Philippines	✓	✓
	Argentina	✓	✓

Source: Researcher

Table number 2 shows that the selected countries all have plans or programs, and policies/legislation/law in terms of electrification of their transportation industry except Armenia.

Table 3. Electrification in Transportation Plans/Programs

COUNTRY	ELECTRIFICATION PLANS/PROGRAMS/POLICIES/LEGISLATION/LAW
DEVELOPED ECONOMIES	
Japan	Japan will endeavor to make all new car sales eco-friendly by the mid-2030s as it joins a growing community of nations determined to slow the globe's carbon emissions (Nikkei, 2020). Currently, Japan leads the world in endorsing NGVs, particularly HVs. Of the 2.5 million new passenger cars sold in 2020, HVs accounted for just over 37%, while PHVs, EVs, and FCVs were each under 1% (0.59%, 0.59%, and 0.03%, respectively), showing the remarkable success of HVs in the Japanese vehicle market (Zhang, 2022). Under the Green Growth Strategy Through Achieving Carbon Neutrality in 2050 formulated by the Ministry of Economy, Trade and Industry in October 2020, the government will take comprehensive measures to make EVs account for all new passenger vehicles sold annually no later than the mid-2030s (Zhang, 2022).
New Zealand	On 5 May 2016, the Government announced its Electric Vehicles Program. This included measures to increase the number of electric vehicles in New Zealand with a goal of reaching approximately 64,000 electric vehicles on our roads by the end of 2021 (Te Manatu Wakak Ministry of Transport, 2021). New Zealand has passed legislation that sets standards for electric vehicles (EVs) to ensure it meets its greenhouse gas (GHG) emissions target of a 41pc reduction by 2030 from 2005 levels.

continued on following page

Table 3. Continued

COUNTRY	ELECTRIFICATION PLANS/PROGRAMS/POLICIES/LEGISLATION/LAW
Norway	Almost all of Norway's power generation (98 percent), comes from renewable hydroelectric power. Norway also has a high share of electric vehicles and by 2025 electric vehicles will likely stand for 100 percent of sales of new vehicles. In 2015 Norway commissioned the world's first electric ferry and fishing boat. By 2020, Norway will also have the world's first electric containership ready for operation, and they have set a goal to electrify all domestic air travel by 2040. Finally, Norway accounts for around 30 percent of global battery-powered ship fleet (Hordnes, 2020). The current Government has decided to keep the incentives for zero-emission cars until the end of 2022. After 2022 the incentives will be revised and adjusted in parallel with the market development (Snarveier, 2016).
United States	The Biden administration is driving plans to increase electric transport across the US. Advances such as replacing all Government vehicles with electric ones have resulted in a new executive order. The President demands all light-duty vehicles acquired by the government to be emission-free by 2027 (Electrive, 2021). The list below (U.S. Department of Energy, n.d.) contains summaries of all Federal laws and incentives related to electricity. Advanced Energy Research Project Grants The Advanced Research Projects Agency - Energy (ARPA-E) was established within the U.S. Department of Energy with the mission to fund projects that will develop transformational technologies that reduce the nation's dependence on foreign energy imports; reduce U.S. energy related emissions, including greenhouse gases; improve energy efficiency across all sectors of the economy; and ensure that the United States maintains its leadership in developing and deploying advanced energy technologies. Advanced Technology Vehicle (ATV) and Alternative Fuel Infrastructure Manufacturing Incentives Through the U.S. Department of Energy's (DOE) Advanced Technology Vehicles Manufacturing Loan Program, manufacturers may be eligible for direct loans for up to 30% of the cost of re-equipping, expanding, or establishing manufacturing facilities in the United States used to produce qualified ATVs, ATV components, or alternative fuel infrastructure, including associated hardware and software. Airport Zero Emission Vehicle (ZEV) and Infrastructure Incentives The Zero Emissions Airport Vehicle and Infrastructure Pilot Program provides funding to airports for up to 50% of the cost to acquire ZEVs and install or modify supporting infrastructure for acquired vehicles. Grant funding must be used for airport-owned, on-road vehicles used exclusively for airport purposes. In 2019, the Ministry of Environment initiated the development and adoption of a legislative provision effectively waiving the 20% Value Added Tax (VAT) on the import and sale of electric vehicles until the end of 2021. Coupled with the Eurasian Economic Union's decision in May 2020 to waive customs taxes on e-vehicles, this move has led to very favorable import and sales conditions for e-vehicles in Armenia, resulting in unprecedented growth for e-vehicle imports and sales over the past year (15,000% growth, or 155-fold in the first half of 2020 as compared to e-vehicles imported in the first half of 2018) (Ghaplanyan, and Grigoryan, 2021).
ECONOMIES IN TRANSITION	

continued on following page

Table 3. Continued

COUNTRY	ELECTRIFICATION PLANS/PROGRAMS/POLICIES/LEGISLATION/LAW
Armenia	Since the IEA's last review in 2014/15, Armenia has developed an Energy Strategy, released in 2021, which calls for up to 1 000 MW of solar PV capacity to be installed by 2030, lifting the share of grid-connected solar to 15% of generation. Armenia is heavily promoting renewables not only to increase energy security but also to meet greenhouse-gas reduction commitments (IEA, 2022).
Albania	The National Strategy for Development and Integration 2015-2020 (NSDI-II)
DEVELOPING ECONOMIES	
Egypt	Egypt will manufacture its first electric vehicle in 2023, President Abdel-Fattah El-Sisi announced on Tuesday during a World Youth Forum (WYF) panel discussion, as part of the state's efforts to transition towards a green economy (Ahram Online, 2022). EGYPT's EVOLVING POLICY FRAMEWORK FOR ELECTRIC VEHICLES The Government of Egypt (GoE) continues to signal a strong determination to increase dependency on electric vehicles (EVs) in Egypt, including by creating an enabling environment for EV local manufacturing and the gradual roll-out of a framework for EVs licensing and operation. So far, the framework has included custom breaks for importing EVs and their local components as well as setting a competitive tariff for electric charging. Efforts are also underway to enhance Egypt's electric charging infrastructure (LYNX Strategic Business Advisors, 2020).
Philippines	The Philippine electric vehicle (e-vehicle) industry aims to have a nation where the use of electric vehicles is highly promoted, encouraged, and supported by its government and society to develop a transportation landscape that is one with the environment, ecologically and economically (GOV. PH, n.d.). The bills regulating the use of vintage and electric vehicles in the country have lapsed into law. President Rodrigo Duterte allowed the two measures to lapse on April 15. Under Republic Act (RA) 11697, a Comprehensive Roadmap for the Electric Vehicle Industry (CREVI) will serve as a national development plan for the electric vehicles (EV) industry to accelerate the development, commercialization, and utilization of EVs in the country. The comprehensive roadmap includes four components – EVs and charging stations; manufacturing; research and development; and human resource development (Gita-Carlos, 2022).
Argentina	Urban Freight Fleet Electrification This scenario assumes a full transition of all vehicles used in "last mile" urban logistics to electric vehicles by 2030 (Furtado, et al., n.d.). Integrated in the National Transportation Plan.

Source: Researcher

Table 3 presents the detailed plans/programs and policies/legislation/laws of the selected countries. It is noticeable that all the countries have incorporated electrification of transportation into their national strategies. However, what is evident is the rate of implementation is certainly faster when it comes to developed economies.

Table 4. Autonomous and Connected Vehicles

COUNTRY		AUTONOMOUS AND CONNECTED VEHICLES	
		PLANS/ PROGRAMS	POLICIES/LEGISLATION/ LAW
DEVELOPED ECONOMIES	Japan	✓	✓
	New Zealand	✓	✓
	Norway	✓	✓
	United States	✓	✓
ECONOMIES IN TRANSITION	Armenia	X	X
	Albania	X	X
DEVELOPING ECONOMIES	Egypt	✓	✓
	Philippines	X	X
	Argentina	X	X

Source: Researcher

Table 5. Autonomous and Connected Vehicles Policies/Legislation/Law

COUNTRY	AUTONOMOUS AND CONNECTED VEHICLES PLANS/PROGRAMS/POLICIES/ LEGISLATION/LAW
DEVELOPED ECONOMIES	
Japan	Japan's government has moved a step closer to allowing fully autonomous driving by adopting a bill that will make driverless cars legal. The move to allow Level 4 autonomous vehicles is an obvious next step after a Honda Legend with Level 3 tech was certified for use on public roads last year by the Japanese Ministry of Land, Infrastructure, Transport and Tourism (Hope, 2022).
New Zealand	There are currently no level 4 or 5 vehicles operating on New Zealand public roads. On the website of the Ministry of Transport, it was emphasized that they are currently developing a long-term insight briefing on the impact of autonomous vehicles operating on New Zealand roads (n.d.).
Norway	Norway is a very early adopter of autonomous driving and tests such as the ones in Drammen raise awareness about the ongoing technological progress (Easymile, 2022). The launched cars in Norway are categorized as Level 4.
United States	The U.S. Department of Transportation (USDOT) is committed to facilitating a new era of transportation innovation and safety and ensuring that our country remains a leader in automation (U.S. Department of transportation, 2016).
ECONOMIES IN TRANSITION	
Armenia	Still in the stage of understanding autonomous and connected cars.
Albania	No information found.
DEVELOPING ECONOMIES	

continued on following page

Table 5. Continued

COUNTRY	AUTONOMOUS AND CONNECTED VEHICLES PLANS/PROGRAMS/POLICIES/ LEGISLATION/LAW
Egypt	Bright skies a technology company, and digital transformation enabler announced today the first prototype release of an Egyptian-made autonomous driving system; a level 4 autonomous driving vehicle that supports highway driving, as specified by the Society of Automotive Engineering (SAE) J3016 (EgyptInnovate, 2021).
Philippines	The Bases Conversion and Development Authority (BCDA) of the Philippines will be providing autonomous vehicle (AV) service in New Clark City during the 2019 Southeast Asian Games (Umali, 2019). Fully autonomous self-driving cars are not yet available in the Philippines. However, in an article published on Philippine Star's website, Seno (2020) reported that there are some key self-driving car technologies available in the Philippines that will hopefully be standard equipment soon: Forward Collision Alert/Warning (FCA) Autonomous Emergency Braking (AEB) Self-Parking (SPS)
Argentina	Not used yet.

Source: Researcher

Table 5 shows the reception of different countries in terms of autonomous and connected vehicles. In Japan, the current move is to allow Level 4 or High Automation wherein the automated driving system undertakes all aspects of the dynamic driving task, even if a human driver does not respond appropriately to a request to intervene. For New Zealand, the country is still reviewing the approval of Level 4 or Level 5 which is Full Automation which is an automated driving system that undertakes all aspects of the dynamic driving tasks in all roadway and environmental conditions. Norway on the other hand has fully embraced automation. In the United States of America, the use of autonomous vehicles as early as 2015.

Moreover, in the same table, Armenia when compared to the developed economies is currently not yet in the stage of reviewing autonomous and connected vehicles. For Albania, no information was found. In contrast to the countries categorized as two economies in transition, Egypt, a developing economy seems to be at par with developed economies when it comes to adapting to the latest innovations in transportation. Both the Philippines and Argentina do not have fully automated vehicles available in the market yet.

Table 6. Artificial Intelligence in Transportation

COUNTRY		ARTIFICIAL INTELLIGENCE	
		PLANS	POLICIES/LEGISLATION/LAW
DEVELOPED ECONOMIES	Japan	✓	✓
	New Zealand	✓	X
	Norway	✓	✓
	United States	✓	✓
ECONOMIES IN TRANSITION	Armenia	X	X
	Albania	X	X
DEVELOPING ECONOMIES	Egypt	X	X
	Philippines	X	X
	Argentina	X	X

Source: Researcher

Table 7. Artificial Intelligence in Transportation Plans

COUNTRY	ARTIFICIAL INTELLIGENCE PLANS/PROGRAMS/POLICIES/LEGISLATION/LAW
DEVELOPED ECONOMIES	
Japan	The Mitsubishi Corporation has set up a joint venture company that will use artificial intelligence (AI) to control on-demand bus services in Japan. Mitsubishi said that the AI would be used to automatically generate routes, in real time, based on passenger requests that are made through a smartphone app (Frangoul, 2019).
New Zealand	Artificial intelligence in general is being talked about but none specifically for transportation. While parts of our public and private sectors are considering the implications of AI and supporting its development in specific areas, there is no whole-of-government or whole-of-nation approach (Institute of Directors, n.d.).
Norway	AI is present in nearly every major industry, whether it is finance, healthcare, or transportation (Rebellion Research, 2022).

continued on following page

Table 7. Continued

COUNTRY	ARTIFICIAL INTELLIGENCE PLANS/PROGRAMS/POLICIES/LEGISLATION/LAW
United States	The U.S. Department of Transportation (U.S. DOT) (2019) is committed to safety and innovation and sees artificial intelligence (AI) as a promising capability to help achieve these aims. While AI has applicability across many sectors of our economy, U.S. DOT's activities focus on the following ways in which it may be applied within transportation: Enabling the safe integration of AI into the transportation system, including as a foundational technology in many automated driving systems and unmanned aircraft systems. U.S. DOT's work in this area also focuses on safe integration of AI into conventional aircraft systems as well as traffic management operations across modes. Adopting and deploying AI-based tools into internal operations, research, and citizen-facing services. U.S. DOT has focused investments in the application of AI into improving the efficiency and effectiveness of internal processes and research, including natural language processing, computer vision, and machine learning-based predictive analytics. Through Research, Data, and Funding Opportunities in each of these areas, U.S. DOT supports interagency initiatives led by the Office of Science and Technology Policy focused on maintaining American leadership in AI.
ECONOMIES IN TRANSITION	
Armenia	AI is being incorporated into industries like tourism. However, no reports were found on the use of AI in transportation.
Albania	Still in the process of research and development.
DEVELOPING ECONOMIES	
Egypt	Still in the process of research and development.
Philippines	Still in the process of research and development.
Argentina	Still in the process of research and development.

Source: Researcher

Japan is known as a technologically advanced country, and it is indeed true. The country is continuously innovating to improve the use of AI in its transportation systems. Developed economies have already set up plans and set up policies that will support the use of AI in transportation systems.

Albania, Egypt, the Philippines and Argentina do not have much funding to spend on research and development (R&D) purpose as this require a sum of money. Due to their status as low-middle-income countries (classification by the World Bank), these countries still considered as low-middle income.

Table 8. Climate Change Mitigation and Adaptation Measures

COUNTRY	CLIMATE CHANGE MITIGATION AND ADAPTATION MEASURES
DEVELOPED ECONOMIES	
Japan	Japan is the sixth-largest greenhouse gas emitter in 2016 and plays an important role to attain the long-term climate goals of the Paris Agreement. One of the key policy issues in Japan's energy and environmental policy arena is the energy system transition to achieve 80% emissions reduction in 2050, a current policy goal set in 2016 (Sugiyama et al., 2018). Japan becomes the newest and largest partner to the next phase of UNDP's Climate Promise, providing significant climate funds for 23 countries and territories on delivering their national climate pledges (UNDP, 2022). Japan's Long-Term Strategy emphasizes discontinuous innovation. But this should not result in postponing what can be done with existing technologies. Innovations to decarbonize society include not only technological advances but also economic and social systems, as well as lifestyle innovations (UNDP, 2022). The international community is looking at Japan to strengthen her efforts to address climate change, such as the issue of coal-fired thermal power plants and raising the ambition of the national goal to mitigate climate change (UNDP, 2022).
New Zealand	The following were taken from New Zealand's Ministry for the Environment (2022). New Zealand has a history of solving challenges through innovation, resourcefulness and determination. Adapting to climate change is no different. A national adaptation plan will help Aotearoa New Zealand minimize damage from a changing climate. This first national climate change risk assessment helped the Government identify where it needed to prioritize action. It is an important step towards making New Zealand more resilient. Outlining the role of central government in helping Aotearoa New Zealand adapt to the impacts of climate change. The likely impacts of climate change on our coastal areas and how central and local government are preparing for these impacts. The role of councils in helping Aotearoa New Zealand prepare for and adapt to climate change and information to assist them in this role. New Zealand needs to be better prepared to adapt to the effects of climate change. To help achieve this, in November 2016 the Government asked a group of technical experts across the public and private sectors to provide advice on how New Zealand could adapt to the effects of climate change.

continued on following page

Table 8. Continued

COUNTRY	CLIMATE CHANGE MITIGATION AND ADAPTATION MEASURES
Norway	The following were drawn Norway's Ministry of Climate and Environment (2021). Climate change adaptation in Norway was prepared and adopted by the Norwegian Parliament. Since the release of Norway's Sixth National Communication in 2014, Norway has passed several milestones in its work related to climate change adaptation, and important progress has been made on local to national administrative levels and across different sectors. Projections of climate change for Norway have been updated, an Official Norwegian Report (NOU) concerning urban storm water management has been published, and the development of guidelines on how to integrate climate change adaptation into municipal planning activities is under way. Capacity building has been strengthened through networks, cooperation and other activities related to climate change adaptation. Climate change adaptation is also integrated into strategies and action plans within and across relevant sectors, such as in the recently adopted White Paper Risk in a Safe and Secure Society (Meld.St. 10 (2016-2017)), where climate change is considered one of the major threats to the Norwegian society.
United States	The United States Environmental Protection Agency have a list of strategies addressing climate change in air, water, waste, and public health (n.d.).
ECONOMIES IN TRANSITION	
Armenia	NATIONAL ACTION PROGRAM OF ADAPTATION TO CLIMATE CHANGE AND THE LIST OF MEASURES FOR 2021-2025 Recognizing the implications of climate change (hereinafter - CC) for its national development, the Government of the Republic of Armenia has and is. taking action to support climate change adaptation (hereinafter - CCA or adaptation) and its planning, with institutional arrangements that address CC already in place, and increased emphasis on CC-related policies and programs. In response to national needs and international commitments, with the support of the GCF-UNDP project "National Adaptation Plan to Advance Medium- and Long-Term Adaptation Planning in Armenia" this National Action Program of Adaptation to Climate Change and the List of Measures for 2021- 2025 (hereinafter - Program) was developed to identify the strategic framework of adaptation and efforts towards national adaptation planning, as well as aiming to guide the adaptation on an ongoing and rolling basis in the future. The Program serves as a roadmap for ensuring the mainstreaming of adaptation and implementation of the NAP in the sector policy, regional and local development, and respective financial planning processes, while providing a reference point for bringing together and coordinating various adaptation planning efforts from different sectors. The strategic approaches and principles, proposed coordination mechanisms in the Program are both contextually relevant and keeping with best practices since CC will impact sectors and communities differently. This approach supports the formulation and implementation of location and context-specific adaptation plans that seek to enhance the adaptive capacity and resilience of sectors and communities within their specific context.

continued on following page

Table 8. Continued

COUNTRY	CLIMATE CHANGE MITIGATION AND ADAPTATION MEASURES
Albania	Albania's main sectoral priorities are (Dibra, Tafaj, and Borde, 2019): Integrated water resource management/flood protection Coastal ecosystems management Climate-resilient irrigation, drainage, and flood protection Sustainable agricultural intensification/adapted farm production Sustainable land-use planning and management/population and settlements Housing/green cities/initiative for municipal adaptation Upgrading civil defense preparedness and disaster risk reduction Climate-proof infrastructure Promoting implementation of adaptation strategies for the health sector
DEVELOPING ECONOMIES	
Egypt	Egypt is showing commitment to the climate agenda as it's finalizing the National Climate Change Strategy to support its 2030 SDG development agenda. And it is also preparing to host the COP27—the 27th Session of the United Nations Climate Change Conference of the Parties (Wes, 2022).

continued on following page

Table 8. Continued

COUNTRY	CLIMATE CHANGE MITIGATION AND ADAPTATION MEASURES
Philippines	The following were taken from the National Climate Change Action Plan 2011 – 2028 (In response to what has essentially become a global crisis, the government has enacted the Climate Change Act (Republic Act 9729) which provides the policy framework with which to systematically address the growing threats on community life and its impact on the environment (Climate Change Commission, n.d.). The national climate change framework strategy has recently been translated into a National Climate Change Action Plan (NCCAP), which prioritizes food security, water sufficiency, ecological and environmental stability, human security, climate-smart industries and services, sustainable energy, and knowledge and capacity development as the strategic direction for 2011 to 2028.
Argentina	Argentina has submitted two National Communications to the United Nations Frameworks Convention on Climate Change (UNFCCC), laying out the actions that the government has already taken and the analytical basis for its policy response to climate change, as well as its commitments to take future actions within an official international framework. Most countries in Latin America have currently submitted only one National Communication, except for Mexico (with 3) and Uruguay (with 2). The First National Communication (1997) established the National GHG Inventory for the year 1990 and 1994 and described the country's vulnerability aspects to climate change. The Revision of the First National Communication submitted to the UNFCCC in 1999 included a new GHG Inventory for the year 1997. The Second National Communication (2007) updated the National GHG Inventory for the year 2000 (agriculture, livestock, land use change and forestry are included in this inventory) and included a revision of the 1990, 1994 and 1997 inventories, identified adaptation and mitigation measures to climate change as well as included vulnerability studies and potential climate scenarios. The National Program for Climate Change Impacts was created with the purpose of developing a national strategy for the better understanding of Argentina's vulnerability to climate change impacts and for the development of adaptation measures to climate change. Among its functions is the implementation of studies related to the global climate change impact on the various climates of Argentina, elaboration of national climate change adaptation measures, coordination of climate change actions with other related national programs such as the one for desertification prevention.

Source: Researcher

Examining the different adaptation and mitigation strategies of the selected countries, one will notice that they are similar. The difference is how these strategies are implemented.

Figure 2. Comparison of Performance of Selected Countries in terms of Adapting Transportation Trends

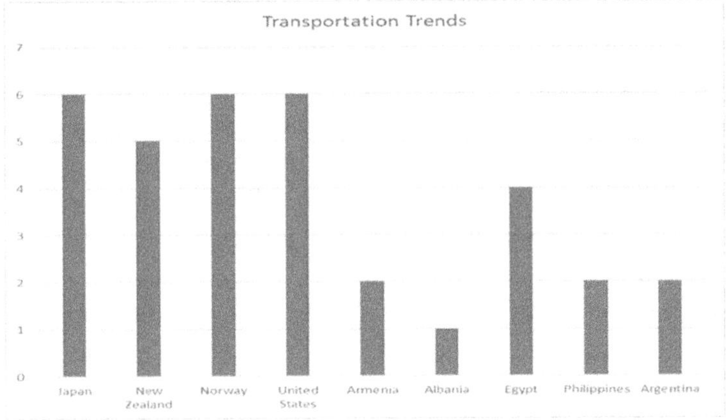

The figure above shows that developed economies fare better in terms of adapting to new transportation trends. This is because they have a high income and a good Human Development Index (HDI) The economies in transition and the developing countries lagged the developed economies.

6: CONCLUSION AND RECOMMENDATIONS

Transportation is undoubtedly an essential sector in the world. The movement of goods and people have led to the growth of the world economy. In terms of transportation trends, one will notice that electrification, automated and connected vehicles, and artificial intelligence aim to make the lives of people more comfortable and mitigate climate change. However, adapting to electrification also comes with challenges like several technological and governance-related challenges.

For autonomous and connected vehicles it is said to have the potential to eliminate human error from driving and increase fuel efficiency by optimizing the route and drive-cycle of the vehicle over the whole projected journey. However, there are also concerns like accidents with automated vehicles, and the reliability and safety of these automated vehicles in different weather and road conditions. There are also ethical issues being faced with life-versus-life situations.

When it comes to artificial intelligence, it is anticipated that AI will contribute to the improvement of human safety, efficiency in traffic, ease of traffic congestion, free driver's time, make parking easier, encourage car- and ridesharing, reduce fuel consumption, and improve air quality. It is also seen to create new jobs, the disappearance of others, and the modification of most. AI can make workers' lives easier by helping workers in the car manufacturing industry for instance, with repetitive, difficult, and dangerous tasks. With this, job positions that might potentially be affected must investigate reskilling and/or upskilling. Moreover, there are also concerns regarding the security of data being collected by AIs. Public acceptance should also be investigated.

The greatest challenge to world transportation will be climate change. With transportation being a major contributor to climate change, the sector must take measures to adapt and mitigate climate change. Climate change on the other hand, also influences the transportation systems and infrastructure.

Lastly, the category of a country influences its ability to adapt to transportation trends. Developed economies are swifter in adapting to these trends. To achieve the intended goals of these developments and innovations, policies and legislation should be present.

With this, the following are the recommendations:

1. For electrification, proper planning of charging locations should be investigated. On top of this, the cost of electricity should also be considered. Although the electrification of vehicles is seen to mitigate climate change, encouraging people to shift to such should be attractive and deemed beneficial.
2. Should the government encourage the shift from fossil fuel vehicles to electric vehicles, programs on the recycling of batteries used by these EVs should also be in place.
3. The government should also have a program that will assist individuals who decide to discontinue using their fossil fuel-run vehicles. Instead of selling these vehicles to other nations, proper disposal or there is a way to recycle these vehicles to minimize waste.
4. If only the steps will be accomplished, then this is the time that the government can investigate encouraging people to purchase electric vehicles. Should the government push through with this, tax incentives for a certain period will be a good tool to entice people.
5. The government should also make sure that the Republic Act (RA) 11697, a Comprehensive Roadmap for the Electric Vehicle Industry (CREVI) will be implemented properly. Control measures should be present to ensure efficient implementation.
6. For autonomous and connected vehicles, the government should study the present road conditions and infrastructure if these are indeed fit for autonomous and connected vehicles.
7. Before adopting ACVs, extensive research should be conducted to ensure the safety of passengers. Moreover, the ethical considerations should be properly addressed.

8. With the emergence of the use of AI in transportation, reskilling and upskilling of affected individuals should be prioritized.
9. If the government decides to adapt to these rising trends, it should make sure that all the stakeholders in the country are properly educated on the benefits and implications of these technological advancements.
10. The government should improve the implementation of the strategies in the National Climate Change Action Plan 2011 – 2028.
11. For the adaptation of trends, and adaptation and mitigation of climate change, cooperation, collaboration, synergy, and convergence are critical and should be practiced both intra agency and interagency.

REFERENCES

Anwar, S., Zia, M. Y. I., Rashid, M., de Rubens, G. Z., & Enevoldsen, P. (2020). Towards Ferry Electrification in the Maritime Sector. *Energies*, *13*(24), 6506. DOI: 10.3390/en13246506

Brake. (n.d.). CONNECTED AND AUTONOMOUS VEHICLES. Retrieved from https://www.brake.org.uk/get-involved/take-action/mybrake/knowledge-centre/vehicles/connected-and-autonomous-vehicles, Nov 20, 2024.

Broom, D. (2021). 4 reasons why electric cars haven't taken off yet. Retrieved from https://www.weforum.org/agenda/2021/07/electric-cars-batteries-fossil-fuel, 08 July.

Climate Change Commission. (n.d.) National Climate Change Action Plan 2011 – 2028. Retrieved from ((https://climate.emb.gov.ph/wp-content/uploads/2016/06/NCCAP-1.pdf

Colonel, M. (2020). Electrification of transportation sector nears tipping point. retrieved from https://www.renewableenergyworld.com/opinion-and-commentary/electrification-of-transportation-sector-nears-tipping-point/#gref, 01 July.

Dibra, L., Tafaj, L., & Borde, A. (2019). Striving to Adapt to Climate Change: Lessons from Albania. Retrieved form https://napglobalnetwork.org/2019/10/striving-to-adapt-to-climate-change-lessons-from-albania/, 29 October.

Easymile. (2022) All Weather Performance for Autonomous Driving On The Road In Norway. Retrieved from https://easymile.com/news/all-weather-performance-autonomous-driving-road-norway, 01 March.

EgyptInnovate. (2021). Brightskies Reveal the 1st Autonomous Driving System in Egypt. Retrieved from https://egyptinnovate.com/en/news/brightskies-reveal-1st-autonomous-driving-system-egypt

Electrive. (2021). US plans to drive transport electrification becoming concrete. Retrieved from https://www.electrive.com/2021/12/09/us-plans-to-drive-transport-electrification-becoming-concrete/#:~:text=The%20Biden%20administration%20is%20driving,be%20emission%2Dfree%20by%202027, 9 December.

European Union. (2019). Artificial intelligence in transport: Current and future developments, opportunities, and challenges. Retrieved from https://www.europarl.europa.eu/RegData/etudes/BRIE/2019/635609/EPRS_BRI(2019)635609_EN.pdf

Fjelstul, J. (25 July 2014). Vehicle electrification: On the "green" road to destination sustainability, *Journal of Destination Marketing & Management, 3*(3), 2014, Pages 137-139, ISSN 2212-571X, . (https://www.sciencedirect.com/science/article/pii/S2212571X14000316)DOI: 10.1016/j.jdmm.2014.06.002

Frangoul, A. (2019). Artificial intelligence is going to control on-demand bus services in Japan. Retrieved from https://www.cnbc.com/2019/03/13/artificial-intelligence-to-control-on-demand-bus-services-in-japan.html#:~:text=The%20Mitsubishi%20Corporation%20has%20set,a%20major%20Japanese%20bus%20operator, 13 March.

Furtado, F., Blanco, J. P., Pritchard, J. P., & Vazano, P. (n.d.). Policy Scenarios for Decarbonising Argentina's Transport System. retrieved from https://www.itf-oecd.org/sites/default/files/docs/policy-scenarios-decarbonising-transport-argentina.pdf

Ghaplanyan, I., & Grigoryan, E. (11 March 2021). The Prospects of Electric Mobility in Armenia. Retrieved from https://evnreport.com/economy/the-prospects-of-electric-mobility-in-armenia-2/

Gita-Carlos, R. A. (2022). Bills regulating vintage, electric vehicles lapse into law. Retrieved from https://www.pna.gov.ph/articles/1173013, 26 April.

GOV.PH. (n.d.). Securing The Future of Philippine Industries. Retrieved from https://industry.gov.ph/industry/e-vehicles/

Hockenos, P. (2018). Europe Takes First Steps in Electrifying World's Shipping Fleets. Retrieved from https://e360.yale.edu/features/europe-takes-first-steps-in-electrifying-worlds-shipping-fleet, 22 February.

Hordnes, E. (2020). RACE TO ELECTRIFICATION – NORWAY IN A POLE POSITION. Retrieved from https://www.swecourbaninsight.com/urban-energy/race-to-electrification-norway-in-a-pole-position/#:~:text=By%202020%2C%20Norway%20will%20also,global%20battery%2Dpowered%20ship%20fleet, 31 August.

Hou, B., Bose, B., & Haran, K. (2020). Powering Electric Aircraft at O'Hare Airport: A Case Study. . Retrieved from https://ieeexplore.ieee.org/document/9281871, 16 December.DOI: 10.1109/PESGM41954.2020.9281871

Hou, B., Bose, S., Mela, L., & Haran, K. (2021). Impact of Aviation Electrification on Airports: Flight Scheduling and Charging. retrieved from https://arxiv.org/pdf/2108.08963.pdf, 20 August.

IEA. (2022). Renewables and energy efficiency can reduce Armenia's dependence on energy imports, IEA policy review says. Retrieved from https://www.iea.org/news/renewables-and-energy-efficiency-can-reduce-armenia-s-dependence-on-energy-imports-iea-policy-review-says, 22 March.

Jarayaman, K. (2019). Third and crucial era of aviation. Retrieved from https://www.dnaindia.com/analysis/column-third-and-crucial-era-of-aviation-2795694, 09 October.

Joshi, N. (2019). How AI Can Transform The Transportation Industry. Retrieved from https://www.forbes.com/sites/cognitiveworld/2019/07/26/how-ai-can-transform-the-transportation-industry/?sh=1a2bf0b84964, 26 July.

Lohmann, G., & Duval, D. T. (2011). Critical Aspects of the Tourism-Transport Relationship. Goodfellow Publishers Ltd 2011. Retrieved from https://www.goodfellowpublishers.com/free_files/fileTransportReview.pdf

LYNX Strategic Business Advisors. (2020). EGYPT's EVOLVING POLICY FRAMEWORK FOR ELECTRIC VEHICLES. Retrieved from https://www.lynxegypt.com/assets/pdfs/EVs-Policy-Regulatory.pdf, November 20.

MacEacheran, M. (2019). Start of aviation's third era? Zero-emission seaplanes may soon be a reality. Retrieved from https://www.business-standard.com/article/companies/start-of-aviation-s-third-era-zero-emission-seaplane-may-soon-be-a-reality-119111601300_1.html, 16 November.

Mammadov, R. (2012). The Importance of Transportation in Tourism Sector. 7th Silk Road International Conference "Challenges and Opportunities of Sustainable Economic Development in Eurasian Countries", May 24, 2012 – May 26, 2012. *SSRN*, *24*(May). https://ssrn.com/abstract=223133

Marr, B. (2022). The 3 Biggest Future Trends In Transportation And Mobility. Retrieved from https://www.forbes.com/sites/bernardmarr/2022/01/20/the-3-biggest-future-trends-in-transportation-and-mobility/?sh=3d84087b3783, 20 January.

Ministry for the Environment. (2022). Adapting to climate change. retrieved from https://environment.govt.nz/what-government-is-doing/areas-of-work/climate-change/adapting-to-climate-change/, 11 April.

Ministry of Climate and Environment. (2021). Norway's first adaptation communication. Retrieved from https://unfccc.int/sites/default/files/resource/Adaptation%20Communication%20Norway.pdf, 05 March.

Mitici, M., Pereira, M., & Oliviero, F. (2022). Electric flight scheduling with battery-charging and battery-swapping opportunities, EURO Journal on Transportation and Logistics, Volume 11, 2022, 100074, ISSN 2192-4376, . (https://www.sciencedirect.com/science/article/pii/S2192437622000036), 16 February.DOI: 10.1016/j.ejtl.2022.100074

O'Neill-Carrillo, E.; Lave, M.; Haines, T. (2021). Systemwide Considerations for Electrification of Transportation in Islands and Remote Locations. *Vehicles*. 3, 498–511., 6 August.DOI: 10.3390/vehicles3030030

Online, A. (2022). Egypt set to manufacture first electric vehicle in 2023, Sisi tells WYF. Retrieved from https://english.ahram.org.eg/NewsContent/1/1235/455019/Egypt/Urban--Transport/Egypt-set-to-manufacture-first-electric-vehicle-in.aspx. 11 January.

Page, S., & Ge, Y. (2009). Transportation and tourism: a symbiotic relationship? In *T. Jamal, & M. Robinson The SAGE handbook of tourism studies* (pp. 372–396). SAGE Publications Ltd., DOI: 10.4135/9780857021076.n21

Pellegrino, F. (2021), "Transport and Tourism Relationship", Grasso, F. and Sergi, B.S. (Ed.) Tourism in the Mediterranean Sea, Emerald Publishing Limited, Bingley, pp. 241-256. DOI: 10.1108/978-1-80043-900-920211017

Rebellion Research. (2022). Norway &. *Artificial Intelligence*, *23*(February). https://www.rebellionresearch.com/norway-artificial-intelligence

Ribeiro, M. A., Gursoy, D., & Chi, O. H. (2022). Customer Acceptance of Autonomous Vehicles in Travel and Tourism. *Journal of Travel Research*, *61*(3), 620–636. DOI: 10.1177/0047287521993578

Rizki, M.; Syahputri, J.; Belgiawan, P.F.; Irawan, M.Z. (2021). Electrifying Tourist Mobility in Bali, Indonesia: Setting the Target and Estimating the CO2 Reduction Based on Stated Choice Experiment. Sustainability 2021, 13, 11656. , 21 October. DOI: 10.3390/su132111656

Rodrigue, J. P. (2020). The Geography of Transport Systems. Retrieved from https://transportgeography.org/contents/applications/tourism-transport/

Royce, R. (n.d.). The future is electric. Retrieved from https://www.rolls-royce.com/products-and-services/electrical.aspx

Schwab, A., Thomas, A., Bennett, J., Robertson, E., & Cary, S. (2021). Electrification of Aircraft: Challenges, Barriers, and Potential Impacts. Golden, CO: National Renewable Energy Laboratory. NREL/TP-6A20-80220. https://www.nrel.gov/docs/fy22osti/80220.pdf

Scuttari, A., & Isetti, G. (2019). E-mobility and Sustainable Tourism Transport in Remote Areas. *Zeitschrift für Tourismuswissenschaft*, *11*(2), 237–256. DOI: 10.1515/tw-2019-0014

Seno, L. (2020). Autonomous Car Technology in PH Cars. Retrieved from https://www.philstar.com/the-freeman/cebu-business/2020/09/14/2042343/autonomous-car-technology-ph-cars, 14 September.

Snarveier. (2016). Norwegian EV policy. Retrieved from https://elbil.no/english/norwegian-ev-policy/, 13 May.

Te Manatu Wakak Ministry of Transport. (July 2021). Electric Vehicles Programme. Retrieved from https://www.transport.govt.nz/area-of-interest/environment-and-climate-change/electric-vehicles-programme/#:~:text=Programme%20announced,by%20the%20end%20of%202021

Umali, T. (2019). The Philippines to use autonomous vehicles during SEA Games. Retrieved from https://opengovasia.com/the-philippines-to-use-autonomous-vehicles-during-sea-games/, 17 July.

United Nations. (1992). United Nations Framework Convention on Climate Change. Retrieved from https://unfccc.int/files/essential_background/background_publications_htmlpdf/application/pdf/conveng.pdf

United Nations. (n.d.). Climate Action. Retrieved from https://www.un.org/en/climatechange/climate-solutions/cities-pollutio

United Nations Economic Commission for Europe. (n.d.). Climate Change and Sustainable Transport. Retrieved from https://unece.org/climate-change-and-sustainable-transport

United Nations New York. (2020). World Economic Situation and Prospects. Retrieved from https://www.un.org/development/desa/dpad/wp-content/uploads/sites/45/WESP2020_Annex.pdf

United States Environmental Protection Agency. (2019). Sources of Greenhouse Gas Emissions. Retrieved from https://www.epa.gov/ghgemissions/sources-greenhouse-gas-emissions

United States Environmental Protection Agency. (n.d.). Strategies for Climate Change Adaptation. Retrieved from https://www.epa.gov/arc-x/strategies-climate-change-adaptation

University of Surrey. (2018, November 7). Autonomous vehicles could shape the future of urban tourism. *ScienceDaily*. Retrieved from www.sciencedaily.com/releases/2018/11/181107111845.htm, 29 December.

U.S. Department of Energy. (n.d.). Electricity Laws and Incentives in Federal. Retrieved from https://afdc.energy.gov/fuels/laws/ELEC?state=US

U.S. Department of Transportation. (2019). U.S. DOT Artificial Intelligence Activities. Retrieved from https://www.transportation.gov/AI, 23 September.

Van Truong, N., & Shimizu, T. (2017). The effect of transportation on tourism promotion: Literature review on application of the Computable General Equilibrium (CGE) *Model. Transportation Research Procedia, 25*, 3096–3115. DOI: 10.1016/j.trpro.2017.05.336

Wes, M. (19 April 2022). Egypt: Acting Against Climate Change for A Healthier, More Prosperous Future. https://www.worldbank.org/en/news/opinion/2022/04/19/-egypt-acting-against-climate-change-for-a-healthier-more-prosperous-future

World Tourism Organization. (2021). *International Tourism Highlights* (2020 Edition). UNWTO., DOI: 10.18111/9789284422456

World Tourism Organization. (2021). UNWTO Secretary-General: 'The Climate Emergency Is a Bigger Threat Than Covid' https://www.unwto.org/news/unwto-secretary-general-the-climate-emergency-is-a-bigger-threat-than-covid, 3 November.

World Tourism Organization. (2022). Tourism Enjoys Strong Start to 2022 while Facing New Uncertainties. Retrieved from https://www.unwto.org/news/tourism-enjoys-strong-start-to-2022-while-facing-new-uncertainties, 25 March.

World Tourism Organization. (n.d). Tourism – an economic and social phenomenon. retrieved from https://www.unwto.org/why-tourism

World Tourism Organization. (n.d). Transforming Tourism for Climate Action. Retrieved from https://www.unwto.org/sustainable-development/climate-action#:~:text=The%20tourism%20sector%20is%20highly,the%20resilience%20of%20the%20sector

World Tourism Organization and International Transport Forum. (2019). Transport-related CO2 Emissions of the Tourism Sector – Modelling Results. UNWTO. Madrid, DOI: DOI: 10.18111/9789284416660

World Tourism Organization and United Nations Development Programme. (2017). *Tourism and the Sustainable Development Goals – Journey to 2030*. UNWTO.

Zhang, R. (2022). Decarbonizing Japan's Transportation Sector. Retrieved from https://www.nbr.org/publication/decarbonizing-japans-transportation-sector-toward-net-zero-emissions/, 12 April.

Zhang, R. and Fujimori, S. (2020). The role of transport electrification in global climate change mitigation scenarios. Environ. Res. Lett. 15 034019, 19 February.

Chapter 2
Mapping the Evolution of Green Energy and Sustainable Mobility:
A Bibliometric Study of Green Energy and Mobility

Animesh Kumar Sharma
https://orcid.org/0000-0002-6673-319X
Lovely Professional University, India

Rahul Sharma
https://orcid.org/0000-0001-8880-7527
Lovely Professional University, India

ABSTRACT

This research study performs a bibliometric analysis of the evolution of green energy and sustainable mobility using the PRISMA technque. The dataset, spanning from 1999 to 2024, was analyzed using VOSviewer and Bibliometrix software to identify key trends, influential authors, and institutional contributions. Finding of the study demonstrate the expanding academic interest in sustainability, reflected in the increasing number of publications over the years. The research is predominantly led by institutions in North America and Europe, with the United States leading publications, followed by the United Kingdom, China, Canada and Australia. Leading journals such as Sustainability Switzerland, Energies, and Sustainable Cities and Society emphasize the interdisciplinary nature of the field. The study brings out research clusters around key authors like B.K. Sovacool and D. Pamucar. Aligning with UN SDG 12, the study emphasizes the AI-driven innovations in responsible consumption, offering insights for policymakers and managers aiming to integrate

DOI: 10.4018/979-8-3373-0882-1.ch002

AI for sustainable consumer engagement.

INTRODUCTION

Recent years have seen major advancements in transportation due to the pressing need to combat climate change, lower air pollution, and develop sustainable urban environments (Shah et al., 2021; Sultana et al., 2019). Electric vehicles, or EVs, have become a key component of a cleaner, greener, and more sustainable urban mobility future (Kovačić et al., 2022; Al-Thani et al., 2022). The migration to electric vehicles is more than just a fad; it is a global movement to change how people commute in urban areas while also protecting the environment (Ruggieri et al., 2021; Mavlutova et al., 2023). Vehicles that run on conventional fossil fuels are no longer viable in cities across the globe due to pollution, crowded roads, and fast urbanization (Kashem et al., 2024; Un-Noor et al., 2017). Their substantial contribution to noise pollution, poor air quality, and greenhouse gas emissions all lowers the standard of living in cities. This has led to a search for solutions to lessen the environmental impact of transportation by governments, corporations, and environmental organizations. The fundamental idea behind these solutions is that electric cars are a greener option than traditional cars with internal combustion engines. EVs are regarded as an essential part of the shift to sustainable urban mobility (Ceder, 2021). Electric cars promise a time when transportation won't worsen the environment because they emit no emissions and can be fuelled by renewable energy sources (Miskolczi et al., 2021; Arsenio et al., 2016). But for EVs to be widely adopted, a comprehensive strategy that incorporates infrastructure development, green energy solutions, and legislative changes to make it easier for EVs to integrate into current urban systems is needed (Das et al., 2020; Singh et al., 2024; Barman et al., 2023).

The Rise of Electric Vehicles

Although electric vehicles have existed for more than a century, their mainstreaming has only occurred in recent decades due to technology improvements and increased environmental concerns (Trencher & Wesseling, 2022). Electric vehicles are becoming more and more popular due to several important variables, all of which are changing the transportation industry (Bohnsack & Pinkse, 2017; Tu & Yang, 2019; Coffman et al., 2017). Environmental consciousness is one of the most important motivators (Mishal et al., 2017; Afsar et al., 2016; Birch et al., 2018). There is an urgent need for sustainable transportation solutions due to growing worries about air pollution, climate change, and the depletion of fossil fuels. Since EVs are thought to be a practical means of lowering carbon emissions and reducing dependency on non-renewable energy sources, they are an essential component of the global effort to stop environmental damage. Technological developments are another important reason driving the increasing popularity of EVs (López et al., 2019; Pelegov & Pontes, 2018). The performance, driving range, and price of electric vehicles have all been greatly improved by the development of high-efficiency batteries, especially lithium-ion technology. Additionally, EV ownership is now more convenient thanks to ongoing advancements in the charging infrastructure, such as fast-charging stations and home-charging options (Zentani et al., 2024; Visaria et al., 2022; Varghese et al., 2024). These developments are hastening the switch to electric automobiles from conventional gasoline-powered ones. The promotion of EV adoption has also been greatly aided by government regulations and incentives (Liu et al., 2021; Zhao et al., 2024). To incentivize manufacturers and consumers to invest in electric vehicles, numerous governments worldwide have implemented tax credits, rebates, reduced registration costs, and subsidies (Hardman et al., 2017; Shafiei et al., 2018). Several nations have set aggressive goals to gradually phase out the sale of new gasoline and diesel cars; some of these countries want to sell only electric vehicles by 2030 or 2040 (Patil et al., 2024; Graham & Brungard, 2021). An atmosphere that is conducive to the widespread adoption of electric mobility is being created by these policy initiatives (Rietmann & Lieven, 2019; Griffiths et al. 2021). Furthermore, EVs are becoming more and more popular because of shifting customer tastes (Morton et al., 2017; Huang et al., 2021). More people are thinking about electric vehicles as a viable and ecological choice as their advantages become more widely known. EVs are less expensive to operate, require less upkeep, and fit in with the global trend toward eco-friendly living (Gulzari et al., 2022; Aithal & Aithal, 2022). Sustainability is becoming a more important consideration for consumers when making purchases, which makes EVs a more alluring choice (Mandys, 2021; Kumar & Alok, 2020; Hui, 2019). The popularity of electric vehicles is also being fueled by businesses' dedication to sustainability (Pardo-Bosch et al., 2021; Jiao & Evans, 2016). Prominent automakers are making significant investments in R&D to create economical, successful EVs and build up charging infrastructure (Gönül et al., 2021; Mastoi et al., 2022). In order to achieve their long-term sustainability objectives, several businesses are also electrifying their fleets (Khan et al., 2018; Yu et al., 2022). The global trend toward electric mobility is being strengthened by this company's transition to clean energy transportation (Okoh et al., 2024; Rani & Jayapragash, 2024). With an emphasis on sustainability, energy efficiency, and less environmental effect, the increasing popularity of EVs represents a fundamental shift in how people view transportation (Zhang et al., 2023; Tundys & Wiśniewski, 2023). It is becoming more and more obvious that electric vehicles are the way of the future as towns, states, and nations continue to adopt them.

Electric Vehicles and Urban Mobility

By drastically lowering air and noise pollution, cutting traffic, and improving the general standard of living in cities, electric vehicles provide a revolutionary answer to urban transportation issues (Elassy et al., 2024; Olabi et al., 2023). Urban regions can transition to a cleaner and more effective transportation system by substituting electric vehicles for conventional internal combustion engine vehicles (Wüstenhagen & Kirschstein, 2024; Alanazi, 2023). Air pollution reduction is one of EVs' main advantages. Electric vehicles reduce carbon dioxide, nitrogen oxides, and particulate matter because they have zero exhaust emissions, in contrast to gasoline and diesel-powered vehicles (Mofolasayo, 2023). This promotes international efforts to tackle climate change and improves the quality of the air. The transition to electric vehicles is essential to creating more sustainable and healthy cities (Pan et al., 2023; Almatar, 2024). Reducing noise pollution is another significant benefit. Compared to conventional vehicles, electric vehicles produce less noise, which makes streets quieter and enhances inhabitants' mental health (Luca et al., 2023; Chawla et al., 2023). Road noise reduction is especially helpful in crowded cities where loud noises have a detrimental effect on day-to-day living (Othman et al., 2024)

When EVs are incorporated into smart city infrastructure, they also help with traffic control. Urban transportation is made more flexible and efficient by autonomous electric mobility, car-sharing services, and sophisticated traffic systems (Alam et al., 2024; Sankar et al., 2024). Furthermore, using electric buses, taxis, and shared mobility options as part of public transportation helps cut down on traffic and fuel expenses (Luo et al., 2023; Sumitkumar et al., 2024). These environmentally friendly transportation choices not only save money on commuting but also help the environment. EVs also promote urban space efficiency. Car-sharing, e-scooters, and e-bikes are examples of shared electric mobility services that might lessen reliance on private vehicle ownership (Corti et al., 2024) and, thus, the need for expansive parking lots. This enables cities to transform urban areas into green spaces, pedestrian-friendly zones, and other projects that are driven by the community (Jarrar & Al-Homoud, 2024).

Research Gap

The integration of green energy solutions and the shift to electric vehicles (EVs) as a fundamental part of urban mobility are topics of great interest worldwide because of their potential to solve social, economic, and environmental issues. Nevertheless, despite the notable advancements, there is still a sizable research vacuum that fully explores the relationship between electric cars, urban mobility, and the uptake of green energy solutions. The following areas account for most of this gap. Given the gaps, it is obvious that a bibliometric study on electric cars, urban mobility, and green energy solutions could make a substantial contribution to bridging these gaps by outlining current research trends, highlighting important areas that require more attention, and suggesting directions for future study.

Research Questions

> **RQ1**: Which countries, regions, and institutions are leading the research in electric vehicles, urban mobility, and green energy solutions?
> **RQ2**: What are the key trends and research themes in the literature on electric vehicles (EVs), urban mobility, and green energy solutions over the past two decades?
> **RQ3:** How has the intersection of electric vehicles, urban mobility, and green energy solutions been explored in academic literature, and what are the key gaps in current research?

These research questions will drive a bibliometric analysis, allowing for a systematic evaluation of existing literature and providing insights into emerging trends, gaps, and prospective topics for future study at the nexus of electric cars, urban mobility, and green energy solutions.

RESEARCH METHOD

This research study uses the PRISMA (Preferred Reporting Items for Systematic Reviews and Meta-Analyses) bibliometric technique to conduct a literature review on electric vehicles (EVs), urban mobility, and green energy solutions (Agrawal et al., 2024). The data for this analysis were obtained from the Scopus database, one of the biggest and most complete collections of peer-reviewed academic literature. Scopus was chosen because of its comprehensive coverage of journals, conference papers, and other academic publications, making it a dependable and relevant source for bibliometric analyses. The bibliometric study's primary purpose is to identify significant trends, research themes, influential authors, major institutions, and geographical patterns related to EVs, urban mobility, and green energy solutions.

Data Collection and Selection Criteria

The initial stage in the research process was to define the search terms and selection criteria for identifying relevant papers. The search query was created using a combination of terms linked to electric vehicles, urban transportation, and green energy solutions. Specific phrases such as "electric vehicles," "urban mobility," "sustainable transportation," "green energy solutions," "renewable energy," and "EV infrastructure" were included to guarantee that the dataset included a wide spectrum of literature on these themes. Boolean operators (AND, OR) were employed to combine keywords and narrow the search. The study's inclusion criteria were based on Scopus-indexed publications, ensuring that the data included high-quality, peer-reviewed literature. Furthermore, the study only included articles written in English, excluding conference abstracts, reports, and non-academic publications. Articles with a focus on the integration of electric vehicles, urban mobility, and green energy solutions, or the intersection of these topics, were included. After conducting the search, the relevant records were downloaded and organized in a structured format for further analysis. The resulting dataset consisted of scholarly articles, conference proceedings, and book chapters that provided insights into various aspects of EVs, urban mobility, and green energy solutions. A total of 306 papers were initially extracted from the scopus database, after which duplicates were removed and articles that did not satisfy the study's criteria were rejected, providing a final sample size of 295 publications to analyse (Figure -1).

Figure 1. PRISMA Framework

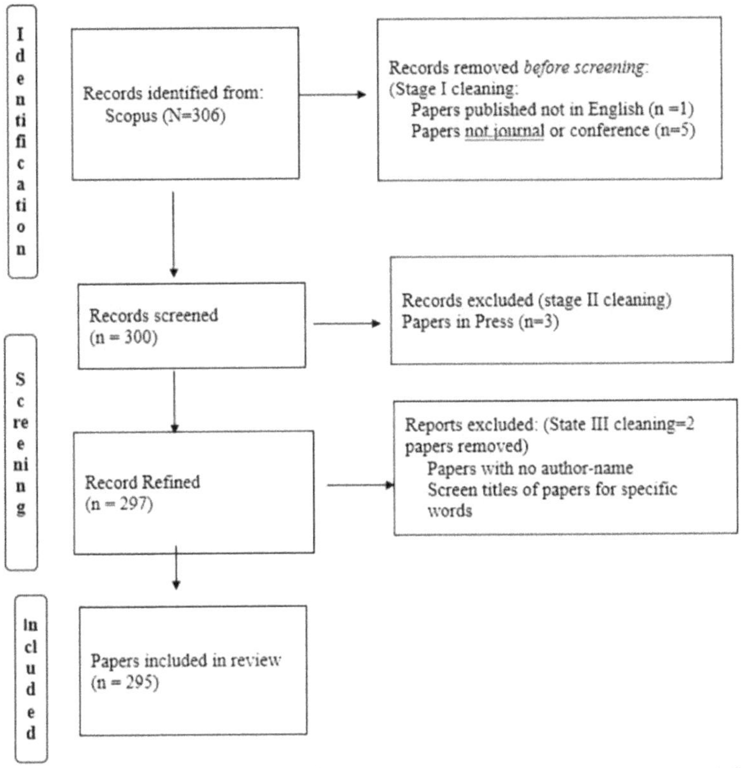

(Source: Authors' own elaboration)

Data Visualization

The data was analyzed using VOSviewer and RStudio's Bibliometric software (Bukar et al., 2023; Ab Rashid, 2023). Several data visualization approaches were used to help people grasp the data better for example, network graphs were utilized to represent co-authorship ties and linkages between prominent writers and institutions. Citation networks were also created to depict the connections between highly cited publications and their contributions to the field. In addition, a timeline graphic was built to demonstrate the progression of significant research themes and trends across time. Keyword co-occurrence networks were developed to highlight the most prevalent topics in literature and their links. These visualisations assisted us in clarifying and intuitively understanding the complicated relationships between the many subject fields, authors, and organisations. The use of data visualization allowed for the identification of critical nodes (e.g., big publications, authors, or institutions) as well as peripheral concerns, which aids in the detection of research gaps and areas requiring more investigation.

RESULTS AND ANALYSIS

Publications

Figure 2 depicts the research study's publications by document type, emphasizing the range of academic outputs. Articles account for most of all publications (115 entries), showing a heavy emphasis on peer-reviewed journal contributions. Books follow shortly behind 100 articles, presenting significant research packaged as individual volumes. There are 56 reviews, which constitute extensive evaluations of current material. Book chapters provide 23 items, indicating joint research efforts in edited volumes. The total number of conference papers is 8, with an emphasis on the dissemination of findings at academic conferences. In addition, the research contains two editorials, one note, and one brief survey, all of which offer succinct scientific contributions. This distribution highlights the many forms in which research findings are presented and distributed.

Figure 2. Publications

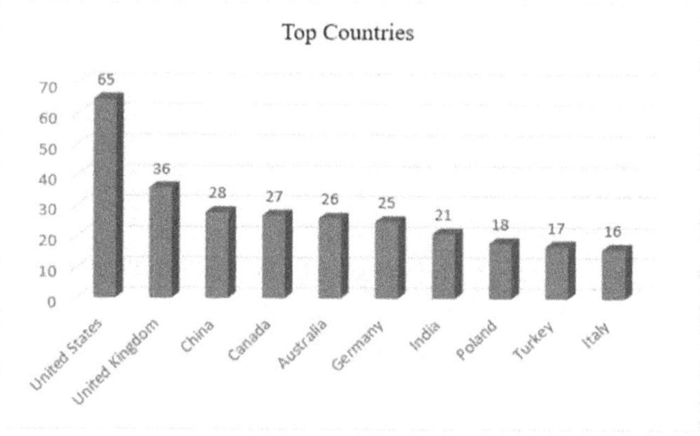

(Source: Authors' own elaboration)

Figure 3 depicts data from a research study on various sorts of open access publications. It classifies OA publications into four types: gold, hybrid gold, bronze, and green. Out of 119 total OA publications, 63 are classified as Gold, which means that research papers are publicly available immediately after publication, generally through author-paid article processing expenses. Hybrid Gold covers 30 periodicals that provide both subscription and open-access alternatives. Bronze OA, which has just five publications, refers to papers that publishers make freely available without an explicit open-access license. Finally, the Green OA model includes 29 publications, which are often self-archived versions put in institutional or topic repositories, giving open access following an embargo period. This classification emphasizes the many ways in which research is made widely accessible.

Figure 3. Open Access Publications

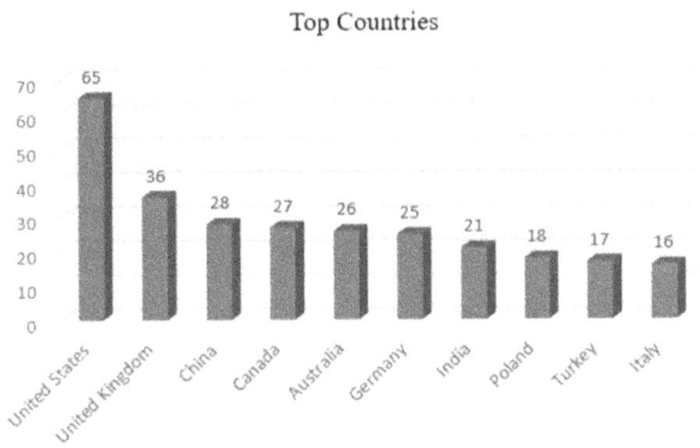

(Source: Authors' own elaboration)

Trends in Publications

Figure 4 depicts publication patterns over time, indicating a large rise in research production in recent years. The most publications were reported in 2024, with 78 contributions, followed by 2023 (46) and 2021 (45), showing a boom in scholarly activity. Between 2022 (41) and 2020 (14), the number of publications remained stable, with changes in previous years. From 2019 to 2015, publishing numbers were moderate, ranging from 4 to 15 each year. Prior to 2015, research output was quite modest, with periodic publications dating back to 1999. The appearance of two publications in 2025 implies that there is continuous scientific effort. Overall, statistics show an increasing trend in research contributions, with a notable rise in recent years, indicating more academic participation and publication attempts.

Figure 4. Trends in Publications

(Source: Authors' own elaboration)

Top Publications Venue

Figure 5 shows the top publication venues for the research study, including the journals with the most research papers published. Sustainability Switzerland leads with 18 publications, suggesting a high emphasis on sustainability-related research. Energies continues with 12 papers that reflect contributions to energy research. Sustainable Cities and Society is ranked third, with ten publications focusing on urban sustainability. Renewable and Sustainable Energy Reviews and the Journal of Cleaner Production have eight and six publications, respectively, indicating a focus on clean energy and sustainable production. IEEE Access has five publications exhibiting involvement in technology and engineering research. Chemical Reviews, Energy Research and Social Science, Science of the Total Environment, and Technological Forecasting and Social Change all have three periodicals that address a variety of multidisciplinary issues. This distribution reveals a substantial preference for sustainability, energy, and environmental research in high-impact publications.

Figure 5. Top Publication Venues

(Source: Authors' own elaboration)

Keywords Occurrences

Figure 6 shows the most used keywords in the research study, which highlight significant theme areas. The most common keyword, Sustainable Development, appears 39 times, showing a significant emphasis on long-term environmental and economic progress. Climate Change follows closely with 33 incidents, emphasizing its importance in the research. Decision Making appears 28 times, indicating a focus on policy and strategic planning. Sustainability and Urban Transportation are both referenced 23 times, indicating a balanced interest in environmental responsibility and mobility solutions. Smart City appears 21 times, highlighting the use of technology into urban design. Artificial intelligence is mentioned 17 times, demonstrating its rising importance in sustainability initiatives. Electric Vehicle (14) and Electric Vehicles (13) combined indicate substantial study toward sustainable transportation. Energy Efficiency (14) appears as a prominent theme, reflecting the study's emphasis on optimizing energy consumption. These terms illustrate the study's multidisciplinary approach, which incorporates sustainability, technology, and decision-making processes (Figures 7 and 8).

Figure 6. Keywords wise Publications

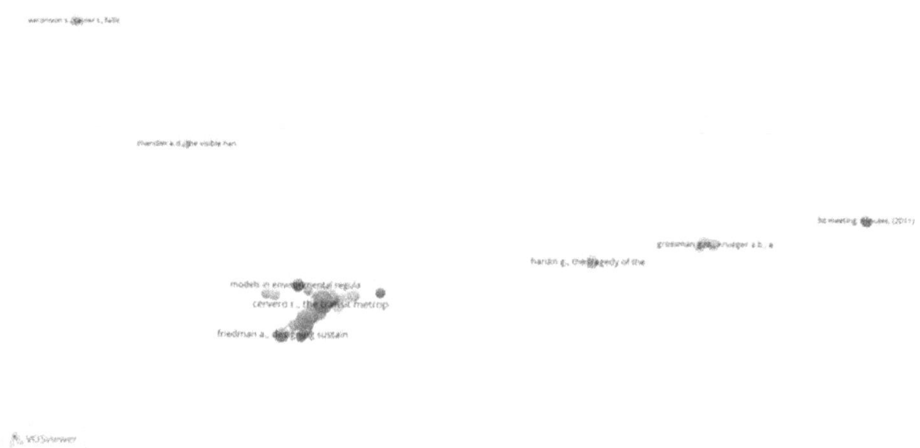

(Source: Authors' own elaboration)

Figure 7. Keywords Occurrences

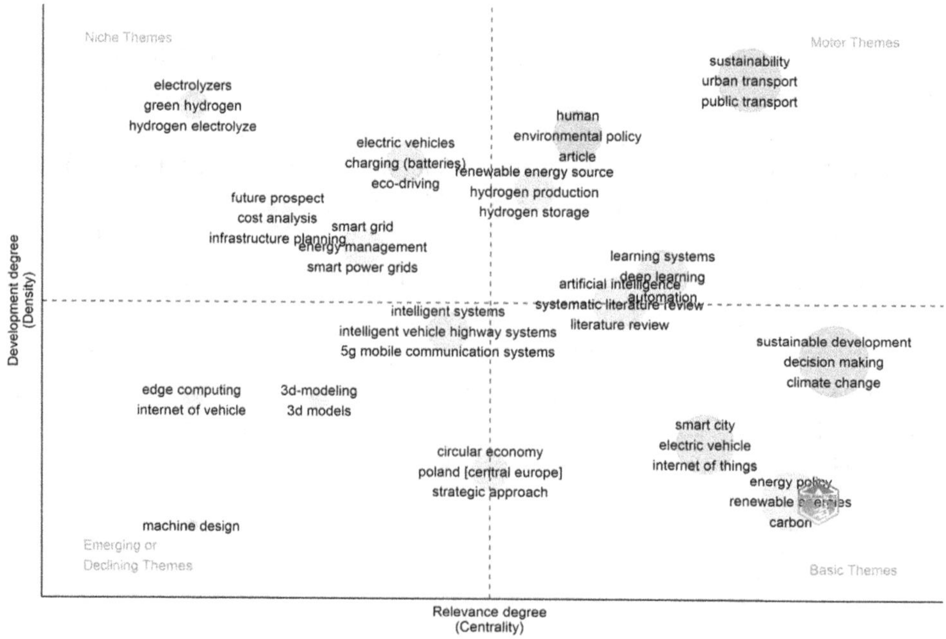

(Source: VOSViewer)

Figure 8. Word cloud

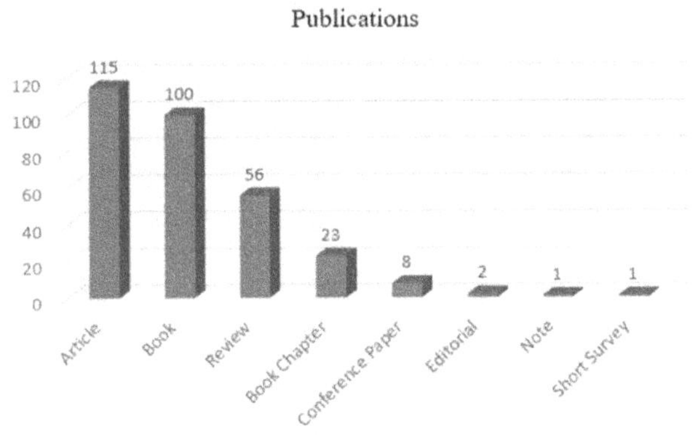

(Source: RStudio's Bibliometrics)

Top Affiliations

Figure 9 depicts the major academic institutions that contributed to the research project, demonstrating a wide range of global affiliations. Leading the list are the University of Waterloo, University of Southampton, Aarhus Universitet, University of Belgrade, and Vilniaus Gedimino Technikos Universitetas, each with five publications, indicating significant contributions from these institutions in the fields of sustainability, technology, and policy research. The University of Žilina, Harvard University, Rheinisch-Westfälische Technische Hochschule Aachen, University of Alberta, and University of Sussex Business School all have four publications, indicating their multidisciplinary research focus. This distribution illustrates the study's multinational collaboration, with contributions from top universities in North America, Europe, and beyond, emphasizing the research themes' global importance.

Figure 9. Top Affiliations

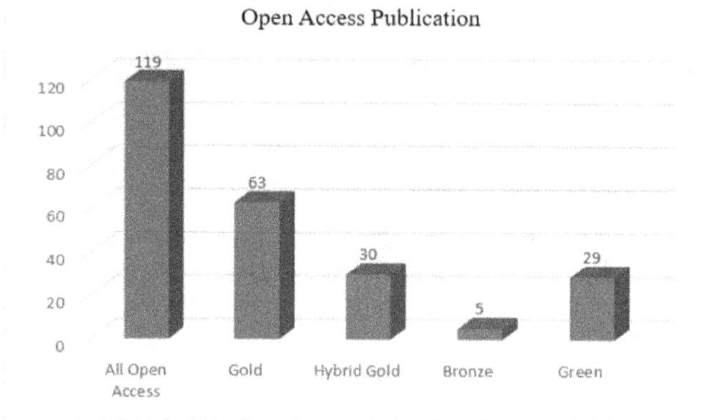

(Source: Authors' own elaboration)

Top Countries

Figure 10 shows the top contributing nations to the research study, demonstrating a broad global presence. The United States leads with 65 publications, reflecting its leading position in research production. The United Kingdom follows with 36 articles, indicating its strong intellectual contributions. China (28), Canada (27), and Australia (26) all take an active role in sustainability, technology, and policy research. Germany (25) also plays an important role, contributing to progress in a variety of scientific sectors. India (21) highlights its expanding research presence, notably in new technologies and sustainable development. Poland (18), Turkey (17), and Italy (16) highlight the study's various international collaborations. This distribution emphasizes the study's worldwide relevance, incorporating opinions from both developed and developing countries.

Figure 10. Top Countries

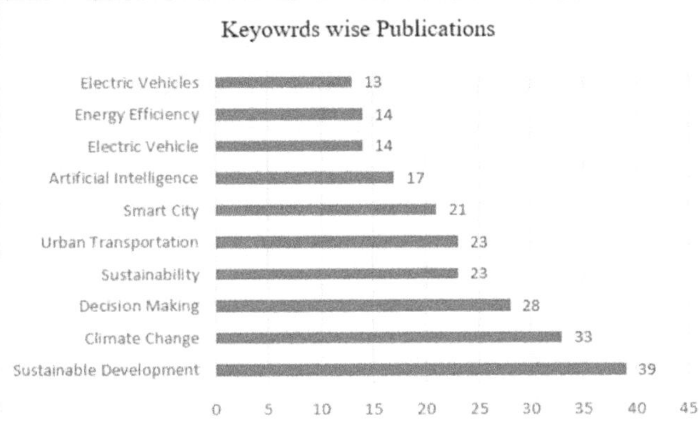

(Source: Authors' own elaboration)

Top Authors

Figure 11 depicts the top contributing authors to the research paper, emphasizing significant scholars advancing the area. Sovacool, B.K. leads with 5 publications, demonstrating a significant effect in sustainability, energy, and policy research. Pamucar, D. follows closely behind with four contributions that demonstrate skill in decision-making frameworks and applied approaches. Several authors, including Brendel, A.B., Deveci, M., Moslem, S., Stević, Ž., and Yigitcanlar, T., each have three publications, displaying their active engagement in multidisciplinary research fields such as smart cities, AI, and urban sustainability. Furthermore, Caban, J., Cao, D., and Cheshmehzangi, A. have each contributed two articles, demonstrating the study's range of experience. This distribution shows a collaborative research effort that spans numerous fields and geographies, enhancing the study's depth and significance.

Figure 11. Top Authors

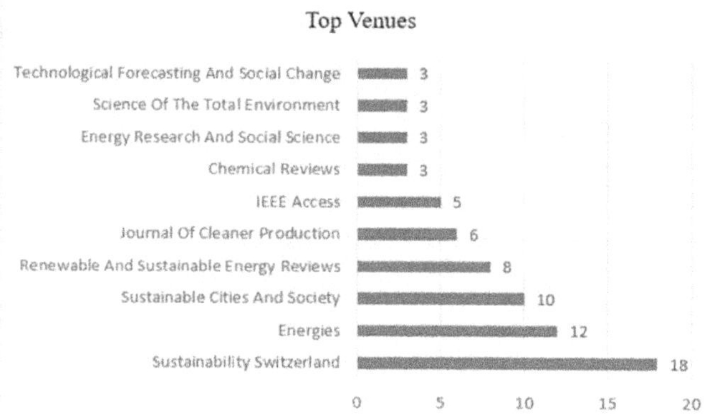

(Source: Authors' own elaboration)

Co-authorship Analysis

Figure 12 depicts a co-authorship network, with each node representing a single author and edges representing cooperation on scholarly articles. Nodes are color-coded, signifying unique groups of regularly co-authoring researchers. The red cluster contains writers such as Deveci, Muhammet, Pamucar, Dragan, and Goksar, Ilgin, who have substantial links, indicating regular cooperation. Similarly, the green cluster, which contains Moslem, Sarbast, Pilla, Francesco, and Szmelter-Jarosz, Agnieszka, indicates another active group, whilst the blue cluster, which includes Simic, Vladimir, Saha, Abhijit, and Senapati, Tapan, has clear co-authorship relationships, albeit less densely connected. Central authors such as Saha, Abhijit (blue cluster) and Moslem, Sarbast (green cluster) play important roles in their respective groups, while Stevic, Jelko, who is located at the junction of many clusters, most likely serves as a link across study fields. Cross-cluster partnerships do exist, such as between Deveci, Muhammet (red) and Saha, Abhijit (blue), although they tend to be less prevalent. Collaboration intensity is strong among closely related writers such as Torkayesh, Ali Ebadi and Zavadskas, Edmundas Kazimieras, whilst others, such as Simic and Vladimir, have fewer links, reflecting restricted cooperation. The entire network has a reasonably dense structure, with well-connected researchers and some isolated nodes, reflecting both cooperation and autonomous study. The existence of many clusters shows a split in study areas, while overlaps indicate multidisciplinary collaboration. Furthermore, the different names imply multinational connections, emphasizing the network's multidisciplinary character.

Figure 12. Co-authorship Network

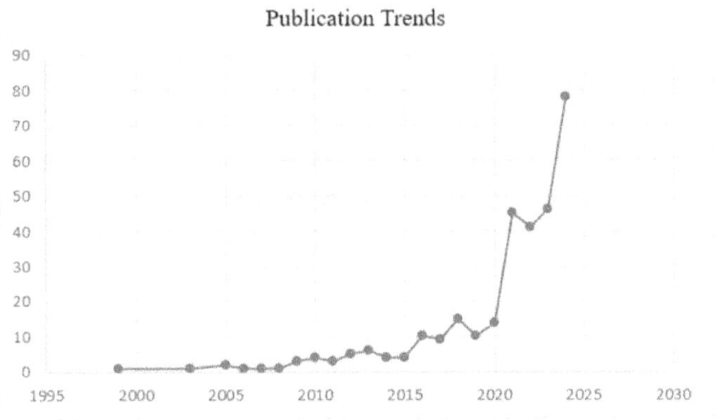

(Source: VOSViewer)

Bibliographic Coupling

Figure 13 pesnts a bibliographic coupling, especially a network or relationship map, depicting relationships between distinct things. The map is made up of colored nodes that represent various entities or concepts, each labeled with text, which might be the names of individuals, organizations, or other components. The colors most likely distinguish these items into various groupings. The lines (edges) linking the nodes represent linkages or interactions, with thickness variations possibly indicating the strength or frequency of these connections. An interesting aspect is the node labeled "Gordon" with the number "22," implying a central entity with several connections, presumably representing influence or prominence in the network. The general layout is either radial or hierarchical, with Gordon serving as the focal point. The basic white backdrop promotes clarity, while a little symbol in the bottom left corner can function as a legend or key, describing the colors and symbols utilized. Although the particular topic is unclear without more information, this map might represent a social, corporate, or scientific collaboration network that is used to analyze complicated interactions and uncover crucial trends in the dataset.

Figure 13. Bibliometric Coupling

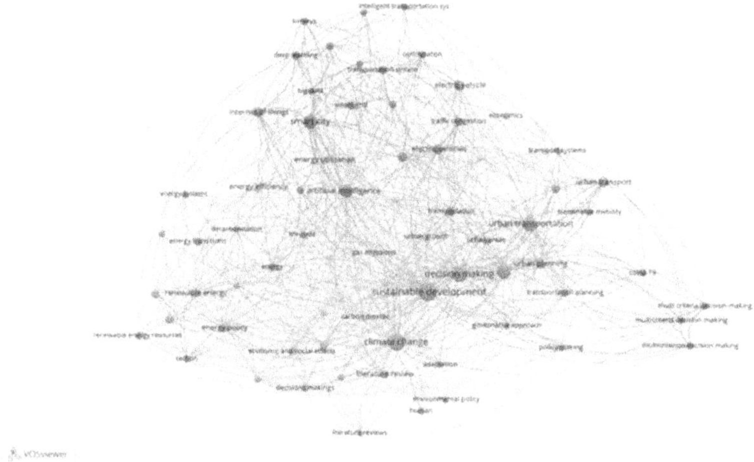

(Source: VOSViewer)

Co-Citation Analysis

Figure 14 depicts a co-citation network created with VOSviewer, demonstrating links between authors or works based on their co-citations in academic literature. Thematic groups are indicated by many clusters of nodes, each labeled with an author's name or publication title. Notable writers include Aaronson S., Cajner T., Fallic, Chandler A.D., Hardin G., and Cervero R., whose works are frequently mentioned together, implying topical ties. One significant cluster focuses on urban studies, transportation, and sustainability, and includes works by Cervero R. and allusions to The Transit Metropolis. Another cluster focuses on environmental laws and models, including resources such as Models in Environmental Regulation and Designing Sustain by Friedman A. The presence of Hardin G. and Grossman G.M. implies debates of social issues like The Tragedy of the Commons and economic theories. The nodes are color-coded, indicating the degree or frequency of co-citation, with some clusters being more closely coupled than others. Proximity between nodes indicates substantial co-citation linkages, as seen by the grouping of Hardin G., Grossman G.M., and Krueger A.B., which indicates frequent joint citations in comparable research areas. The inclusion of major books such as The Transit Metropolis and The Tragedy of the Commons emphasizes their significance in urban studies, environmental policy, and sustainability discourse. Overall, this network visualization successfully depicts the linkages between prominent academic contributions from many study disciplines.

Figure 14. Co-citation Network

(Source: VOSViewer)

Thematic Map

Figure 15 depicts a thematic map that divides themes into two categories: "Nice Themes" and "Motor Themes," with each indicating a distinct area of concentration. "Nice Themes" focuses on sustainable and innovative technologies such as electric cars, hydrogen fuel cells, battery technology, smart grids, renewable energy, the circular economy, and design for sustainability, highlighting advances in eco-friendly solutions and energy efficiency. In contrast, "Motor Themes" focuses on traditional and emerging automotive topics such as public transportation, road safety, traffic management, autonomous driving, connected vehicles, vehicle-to-grid integration, and fuel cell vehicles, indicating a preference for transportation infrastructure and vehicle technology. The existence of a "Relevance Axis" and a "Central Axis" implies a categorization based on technical reliance or thematic relevance. This map is a strategic visualization tool that provides insights into the interrelationships of many study topics and aids in planning, analysis, and decision-making in the sustainability and transportation disciplines.

Figure 15. Thematic Map

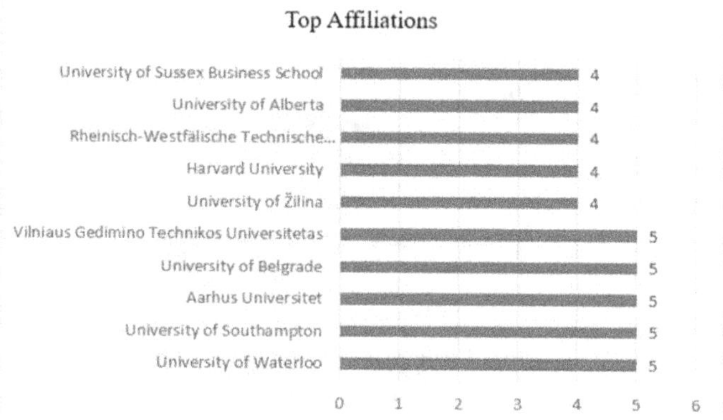

(Source: RStudio's Bibliometrics)

DISCUSSIONS

This research on green energy, electric cars, and sustainable urban transportation sheds light on the progress of these fields. This section analyzes the results' implications, the present research environment, extant gaps, and possible future directions.

Research Output and Trends

The rising number of publications over time demonstrates academic interest in sustainability-related themes. The increase in research production, particularly after 2020, coincides with worldwide policy moves towards sustainability, electrification, and decarbonization. The predominance of journal articles (115) shows a strong emphasis on peer-reviewed, empirical research, whilst the large number of books (100) reflects theoretical expansion and consolidation. The existence of review articles (56) indicates a need for synthesis of current information, which aids in the identification of research gaps and future objectives. The increase in open-access publications, which includes 63 gold OA contributions, demonstrates the growing commitment to knowledge dissemination and accessibility. However, the low number of green and bronze OA publications implies that institutional and self-archiving efforts might be improved.

Geographic and Institutional Leadership

The study is mostly spearheaded by institutions in North America and Europe, with the United States leading the way (65 publications), followed by the United Kingdom (36), China (28), Canada (27), and Australia (26). This supremacy is a result of significant government and commercial expenditures in green energy and electric mobility research in these regions. Universities such as the University of Waterloo, the University of Southampton, and Aarhus Universitet emerge as major contributors to intellectual debate. The participation of several German and Indian institutes demonstrates their expanding significance in sustainable transportation and energy research. Despite this, developing countries' contributions are disproportionately underrepresented, notably in Africa and Latin America. Given their distinct urban mobility issues and energy transition requirements, more study from these locations might give useful insights.

Key Research Themes and Topics

Keyword analysis suggests that "Sustainable Development," "Climate Change," and "Decision Making" are key subjects in the literature, occurring 39, 33, and 28 times, respectively. This suggests a significant link between environmental policy, governance, and technical innovation. The frequent use of "Urban Transportation" (23) and "Smart City" (21) illustrates the convergence of mobility solutions with smart infrastructure improvements. The prominent appearance of "Artificial Intelligence" (17) in keyword analysis demonstrates its growing importance in improving sustainable energy consumption and mobility systems. The comparatively low frequency of phrases such as "Policy Implementation" and "Consumer Adoption" suggests that, while technology breakthroughs are well-documented, study into regulatory frameworks and social acceptability remains a topic for additional investigation.

Leading Publication Venues

The importance of journals such as Sustainability Switzerland (18 publications), Energies (12), and Sustainable Cities and Society (10) demonstrates the multidisciplinary character of research. The inclusion of IEEE Access, technical Forecasting, and Social Change emphasizes the relationship between technical breakthroughs and sustainable mobility trends. While these journals provide extensive coverage of the area, the comparatively low number of articles in policy-oriented journals indicates that more involvement in policymaking and regulatory frameworks is required. Strengthening links between technology research and policy studies might boost the effectiveness of sustainability projects.

Collaborative Research Networks

Co-authorship analysis indicates unique research clusters, with significant cooperation networks concentrated on top authors like B.K. Sovacool and D. Pamucar. These clusters represent well-established research communities that concentrate on certain areas of sustainable mobility and energy. However, the scarcity of cross-cluster interactions emphasizes the need for multidisciplinary partnerships that include technical, economic, and policy views. Authors like Jelko Stevic, who cross several clusters, play an important role in promoting interdisciplinary conversation. Encouraging more such collaborations might help to encourage comprehensive research approaches that consider the environmental, social, and economic aspects of sustainability.

Alignment with SDG 12: Responsible Consumption and Production

The study's findings are consistent with SDG 12, stressing how AI-driven technologies may promote responsible consumerism. Businesses may use machine learning algorithms and data-driven insights to prevent overproduction, material waste, and promote eco-friendly alternatives. AI-powered recommendation systems encourage customers to choose sustainable products by making individualized recommendations based on ethical sourcing, carbon footprint analysis, and recyclability indicators. Furthermore, AI applications in supply chain optimization help to reduce energy consumption and mitigate environmental consequences, which reinforces SDG 12's goals.

CONCLUSION

The study provides an in-depth look at the research landscape in green energy and sustainable transportation. The increasing number of publications, significant authors, and prevalent research subjects suggest that sustainability is becoming more important in academic discourse. To make significant progress, future research must address policy concerns, incorporate varied regional viewpoints, and develop multidisciplinary cooperation. Bridging these gaps will be critical for expediting the worldwide shift to sustainable urban mobility and green energy solutions. This report emphasizes AI's disruptive influence on consumer behavior and sustainable consumption, while also identifying potential and difficulties. While AI-powered devices improve efficiency, decrease waste, and encourage responsible consumption, they also raise ethical questions about data privacy and prejudice. The findings highlight the need of taking a balanced strategy that combines technology improvements with ethical concerns when aligning AI applications with sustainability goals. Policymakers and corporations may use responsible AI deployment to support sustainable consumer choices while reducing unanticipated negative outcomes.

IMPLICATIONS

Theoretical Implications

This study adds to the expanding body of knowledge on AI-driven consumer behavior and sustainable consumption by combining technology adoption theories with sustainability frameworks. It demonstrates how AI improves decision-making processes, increases efficiency, and encourages environmentally beneficial purchasing patterns. It also emphasizes the need to improve existing theoretical models by including ethical issues, such as data privacy and algorithmic transparency, into AI-driven customer involvement. By bridging the gap between AI innovation and sustainable consumption, this study lays the groundwork for future research on AI's long-term influence on consumer sustainability habits.

Managerial Implications

This study's findings provide useful insights for managers and policymakers looking to use AI to drive long-term customer engagement. Businesses may use AI-driven data to improve product suggestions, streamline supply chains, and minimize waste, all of which correlate with sustainability goals. Furthermore, AI can help with targeted consumer education on eco-friendly options, promoting responsible consumption behaviors. To earn customer confidence, firms must address ethical concerns including data security and algorithmic transparency. Managers must incorporate AI responsibly, ensuring that technical breakthroughs are aligned with both commercial objectives and sustainability goals.

Practical Implications

The study's findings have important practical ramifications for firms, politicians, and consumers. AI-powered technologies can assist customers in making educated selections by giving real-time data on product sustainability, energy efficiency, and environmental effect. Policymakers may create regulatory frameworks to guarantee that AI applications correspond with sustainability objectives while also preserving consumer rights. Businesses may also use AI into their operations to increase efficiency, minimize waste, and encourage environmentally friendly options. By carefully using AI, all stakeholders may contribute to a more sustainable future while taking use of technical breakthroughs for economic and environmental advantages.

LIMITATIONS AND FUTURE RESEARCH DIRECTIONS

While this study sheds light on the function of artificial intelligence in changing consumer behavior and sustainable consumption, it is not without limits. The scopus database was utilized in this investigation. Another database might be studied in the future. This study relies mostly on secondary data and theoretical frameworks, which may restrict empirical validity. To bolster the conclusions, future study might use primary data collected via questionnaires or trials. This study does not take into consideration geographical and cultural differences in AI adoption, which may affect consumer behavior differently across markets. Future research should investigate cross-cultural viewpoints to improve the results' generalizability. Finally, as AI technologies advance, further study is required to analyze their long-term influence on sustainability and consumer decision-making, particularly in emerging economies.

REFERNCES

Ab Rashid, M. F. (2023). How to conduct a bibliometric analysis using R packages: A comprehensive guidelines. *Journal of Tourism. Hospitality and Culinary Arts, 15*(1), 24–39.

Afsar, B., Badir, Y., & Kiani, U. S. (2016). Linking spiritual leadership and employee pro-environmental behavior: The influence of workplace spirituality, intrinsic motivation, and environmental passion. *Journal of Environmental Psychology, 45*, 79–88. DOI: 10.1016/j.jenvp.2015.11.011

Agrawal, S., Oza, P., Kakkar, R., Tanwar, S., Jetani, V., Undhad, J., & Singh, A. (2024). Analysis and recommendation system-based on PRISMA checklist to write systematic review. *Assessing Writing, 61*, 100866. DOI: 10.1016/j.asw.2024.100866

Aithal, P. S., & Aithal, S. (2022). Opportunities and Challenges for Green and Eco-Friendly Nanotechnology in Twenty-First Century. *Sustainable nanotechnology: Strategies, products, and applications*, 31-50. DOI: 10.1002/9781119650294.ch3

Al-Thani, H., Koç, M., Isaifan, R. J., & Bicer, Y. (2022). A review of the integrated renewable energy systems for sustainable urban mobility. *Sustainability (Basel), 14*(17), 10517. DOI: 10.3390/su141710517

Alam, T., Gupta, R., Nasurudeen Ahamed, N., Ullah, A., & Almaghthwi, A. (2024). Smart mobility adoption in sustainable smart cities to establish a growing ecosystem: Challenges and opportunities. *MRS Energy & Sustainability : a Review Journal, 11*(2), 304–316. DOI: 10.1557/s43581-024-00092-4

Alanazi, F. (2023). Electric vehicles: Benefits, challenges, and potential solutions for widespread adaptation. *Applied Sciences (Basel, Switzerland), 13*(10), 6016. DOI: 10.3390/app13106016

Almatar, K. M. (2024). Increasing electric vehicles infrastructure in urban areas for efficiently employing renewable energy. *Environment, Development and Sustainability, 26*(10), 26183–26204. DOI: 10.1007/s10668-023-03723-x

Arsenio, E., Martens, K., & Di Ciommo, F. (2016). Sustainable urban mobility plans: Bridging climate change and equity targets? *Research in Transportation Economics, 55*, 30–39. DOI: 10.1016/j.retrec.2016.04.008

Barman, P., Dutta, L., Bordoloi, S., Kalita, A., Buragohain, P., Bharali, S., & Azzopardi, B. (2023). Renewable energy integration with electric vehicle technology: A review of the existing smart charging approaches. *Renewable & Sustainable Energy Reviews, 183*, 113518. DOI: 10.1016/j.rser.2023.113518

Birch, D., Memery, J., & Kanakaratne, M. D. S. (2018). The mindful consumer: Balancing egoistic and altruistic motivations to purchase local food. *Journal of Retailing and Consumer Services, 40*, 221–228. DOI: 10.1016/j.jretconser.2017.10.013

Bohnsack, R., & Pinkse, J. (2017). Value propositions for disruptive technologies: Reconfiguration tactics in the case of electric vehicles. *California Management Review, 59*(4), 79–96. DOI: 10.1177/0008125617717711

Bukar, U. A., Sayeed, M. S., Razak, S. F. A., Yogarayan, S., Amodu, O. A., & Mahmood, R. A. R. (2023). A method for analyzing text using VOSviewer. *MethodsX, 11*, 102339. DOI: 10.1016/j.mex.2023.102339 PMID: 37693657

Ceder, A. (2021). Urban mobility and public transport: Future perspectives and review. *International Journal of Urban Sciences, 25*(4), 455–479. DOI: 10.1080/12265934.2020.1799846

Chawla, U., Mohnot, R., Mishra, V., Singh, H. V., & Singh, A. K. (2023). Factors influencing customer preference and adoption of electric vehicles in India: A journey towards more sustainable transportation. *Sustainability (Basel), 15*(8), 7020. DOI: 10.3390/su15087020

Coffman, M., Bernstein, P., & Wee, S. (2017). Electric vehicles revisited: A review of factors that affect adoption. *Transport Reviews, 37*(1), 79–93. DOI: 10.1080/01441647.2016.1217282

Corti, F., Iacono, S. D., Astolfi, D., Pasetti, M., Vasile, A., Reatti, A., & Flammini, A. (2024). A comprehensive review of charging infrastructure for Electric Micromobility Vehicles: Technologies and challenges. *Energy Reports, 12*, 545–567. DOI: 10.1016/j.egyr.2024.06.026

Das, H. S., Rahman, M. M., Li, S., & Tan, C. W. (2020). Electric vehicles standards, charging infrastructure, and impact on grid integration: A technological review. *Renewable & Sustainable Energy Reviews, 120*, 109618. DOI: 10.1016/j.rser.2019.109618

Elassy, M., Al-Hattab, M., Takruri, M., & Badawi, S. (2024). Intelligent transportation systems for sustainable smart cities. *Transportation Engineering, 100252*, 100252. Advance online publication. DOI: 10.1016/j.treng.2024.100252

Gönül, Ö., Duman, A. C., & Güler, Ö. (2021). Electric vehicles and charging infrastructure in Turkey: An overview. *Renewable & Sustainable Energy Reviews, 143*, 110913. DOI: 10.1016/j.rser.2021.110913

Graham, J. D., & Brungard, E. (2021). Consumer adoption of plug-in electric vehicles in selected countries. *Future Transportation, 1*(2), 303–325. DOI: 10.3390/futuretransp1020018

Griffiths, S., Del Rio, D. F., & Sovacool, B. (2021). Policy mixes to achieve sustainable mobility after the COVID-19 crisis. *Renewable & Sustainable Energy Reviews, 143*, 110919. DOI: 10.1016/j.rser.2021.110919 PMID: 35702384

Gulzari, A., Wang, Y., & Prybutok, V. (2022). A green experience with eco-friendly cars: A young consumer electric vehicle rental behavioral model. *Journal of Retailing and Consumer Services, 65*, 102877. DOI: 10.1016/j.jretconser.2021.102877

Hardman, S., Chandan, A., Tal, G., & Turrentine, T. (2017). The effectiveness of financial purchase incentives for battery electric vehicles–A review of the evidence. *Renewable & Sustainable Energy Reviews, 80*, 1100–1111. DOI: 10.1016/j.rser.2017.05.255

Huang, Y., Qian, L., Soopramanien, D., & Tyfield, D. (2021). Buy, lease, or share? Consumer preferences for innovative business models in the market for electric vehicles. *Technological Forecasting and Social Change, 166*, 120639. DOI: 10.1016/j.techfore.2021.120639

Hui, A. (2019). Understanding the positioning of "the electric vehicle consumer": Variations in interdisciplinary discourses and their implications for sustainable mobility systems. *Applied Mobilities, 4*(1), 49–65. DOI: 10.1080/23800127.2017.1380977

Jarrar, O. M., & Al-Homoud, M. (2024). Sustainable urban development in Riyadh: A projected model for walkability. *International Journal of Urban Sustainable Development, 16*(1), 398–421. DOI: 10.1080/19463138.2024.2421163

Jiao, N., & Evans, S. (2016). Business models for sustainability: The case of second-life electric vehicle batteries. *Procedia CIRP, 40*, 250–255. DOI: 10.1016/j.procir.2016.01.114

Kashem, M. A., Shamsuddoha, M., & Nasir, T. (2024). Sustainable Transportation Solutions for Intelligent Mobility: A Focus on Renewable Energy and Technological Advancements for Electric Vehicles (EVs) and Flying Cars. *Future Transportation, 4*(3), 874–890. DOI: 10.3390/futuretransp4030042

Khan, W., Ahmad, A., Ahmad, F., & Saad Alam, M. (2018). A comprehensive review of fast charging infrastructure for electric vehicles. *Smart Science, 6*(3), 256–270. DOI: 10.1080/23080477.2018.1437323

Kovačić, M., Mutavdžija, M., & Buntak, K. (2022). New paradigm of sustainable urban mobility: Electric and autonomous vehicles—A review and bibliometric analysis. *Sustainability (Basel), 14*(15), 9525. DOI: 10.3390/su14159525

Kumar, R. R., & Alok, K. (2020). Adoption of electric vehicle: A literature review and prospects for sustainability. *Journal of Cleaner Production, 253*, 119911. DOI: 10.1016/j.jclepro.2019.119911

Liu, X., Sun, X., Zheng, H., & Huang, D. (2021). Do policy incentives drive electric vehicle adoption? Evidence from China. *Transportation Research Part A, Policy and Practice, 150*, 49–62. DOI: 10.1016/j.tra.2021.05.013

López, I., Ibarra, E., Matallana, A., Andreu, J., & Kortabarria, I. (2019). Next generation electric drives for HEV/EV propulsion systems: Technology, trends and challenges. *Renewable & Sustainable Energy Reviews, 114*, 109336. DOI: 10.1016/j.rser.2019.109336

Luca, O., Andrei, L., Iacoboaea, C., & Gaman, F. (2023). Unveiling the hidden effects of automated vehicles on "do no significant harm" components. *Sustainability (Basel), 15*(14), 11265. DOI: 10.3390/su151411265

Luo, H., Chahine, R., Gkritza, K., & Cai, H. (2023). What motivates the use of shared mobility systems and their integration with public transit? Evidence from a choice experiment study. *Transportation Research Part C, Emerging Technologies, 155*, 104286. DOI: 10.1016/j.trc.2023.104286

Mandys, F. (2021). Electric vehicles and consumer choices. *Renewable & Sustainable Energy Reviews, 142*, 110874. DOI: 10.1016/j.rser.2021.110874

Mastoi, M. S., Zhuang, S., Munir, H. M., Haris, M., Hassan, M., Usman, M., Bukhari, S. S. H., & Ro, J. S. (2022). An in-depth analysis of electric vehicle charging station infrastructure, policy implications, and future trends. *Energy Reports, 8*, 11504–11529. DOI: 10.1016/j.egyr.2022.09.011

Mavlutova, I., Atstaja, D., Grasis, J., Kuzmina, J., Uvarova, I., & Roga, D. (2023). Urban transportation concept and sustainable urban mobility in smart cities: A review. *Energies, 16*(8), 3585. DOI: 10.3390/en16083585

Mishal, A., Dubey, R., Gupta, O. K., & Luo, Z. (2017). Dynamics of environmental consciousness and green purchase behaviour: An empirical study. *International Journal of Climate Change Strategies and Management, 9*(5), 682–706. DOI: 10.1108/IJCCSM-11-2016-0168

Miskolczi, M., Földes, D., Munkacsy, A., & Jaszberenyi, M. (2021). Urban mobility scenarios until the 2030s. *Sustainable Cities and Society*, *72*, 103029. DOI: 10.1016/j.scs.2021.103029

Mofolasayo, A. (2023). Assessing and Managing the Direct and Indirect Emissions from Electric and Fossil-Powered Vehicles. *Sustainability (Basel)*, *15*(2), 1138. DOI: 10.3390/su15021138

Morton, C., Anable, J., & Nelson, J. D. (2017). Consumer structure in the emerging market for electric vehicles: Identifying market segments using cluster analysis. *International Journal of Sustainable Transportation*, *11*(6), 443–459. DOI: 10.1080/15568318.2016.1266533

Okoh, A. S., & Onuoha, M. C. (2024). Immediate and future challenges of using electric vehicles for promoting energy efficiency in Africa's clean energy transition. *Global Environmental Change*, *84*, 102789. DOI: 10.1016/j.gloenvcha.2023.102789

Olabi, A. G., Wilberforce, T., Obaideen, K., Sayed, E. T., Shehata, N., Alami, A. H., & Abdelkareem, M. A. (2023). Micromobility: Progress, benefits, challenges, policy and regulations, energy sources and storage, and its role in achieving sustainable development goals. *International Journal of Thermofluids*, *17*, 100292. DOI: 10.1016/j.ijft.2023.100292

Othman, E., Cibilić, I., Poslončec-Petrić, V., & Saadallah, D. (2024). Investigating Noise Mapping in Cities to Associate Noise Levels with Sources of Noise Using Crowdsourcing Applications. *Urban Science (Basel, Switzerland)*, *8*(1), 13. DOI: 10.3390/urbansci8010013

Pan, S., Yu, W., Fulton, L. M., Jung, J., Choi, Y., & Gao, H. O. (2023). Impacts of the large-scale use of passenger electric vehicles on public health in 30 US. metropolitan areas. *Renewable & Sustainable Energy Reviews*, *173*, 113100. DOI: 10.1016/j.rser.2022.113100

Pardo-Bosch, F., Pujadas, P., Morton, C., & Cervera, C. (2021). Sustainable deployment of an electric vehicle public charging infrastructure network from a city business model perspective. *Sustainable Cities and Society*, *71*, 102957. DOI: 10.1016/j.scs.2021.102957

Patil, G., Pode, G., Diouf, B., & Pode, R. (2024). Sustainable Decarbonization of Road Transport: Policies, Current Status, and Challenges of Electric Vehicles. *Sustainability (Basel)*, *16*(18), 8058. DOI: 10.3390/su16188058

Pelegov, D. V., & Pontes, J. (2018). Main drivers of battery industry changes: Electric vehicles—A market overview. *Batteries*, *4*(4), 65. DOI: 10.3390/batteries4040065

Rani, S., & Jayapragash, R. (2024). Review on electric mobility: Trends, challenges and opportunities. *Results in Engineering, 102631*, 102631. Advance online publication. DOI: 10.1016/j.rineng.2024.102631

Rietmann, N., & Lieven, T. (2019). How policy measures succeeded to promote electric mobility–Worldwide review and outlook. *Journal of Cleaner Production, 206*, 66–75. DOI: 10.1016/j.jclepro.2018.09.121

Ruggieri, R., Ruggeri, M., Vinci, G., & Poponi, S. (2021). Electric mobility in a smart city: European overview. *Energies, 14*(2), 315. DOI: 10.3390/en14020315

Sankar, M., Pachiyappan, S., Arjun, B. S., & Srivastava, A. (2024). Assessing the Impact of Smart City Technologies on Sustainable Urban Mobility in Developing Economies. In *Smart Cities and Circular Economy* (pp. 193–214). Emerald Publishing Limited., DOI: 10.1108/978-1-83797-957-820241018

Shafiei, E., Davidsdottir, B., Fazeli, R., Leaver, J., Stefansson, H., & Asgeirsson, E. I. (2018). Macroeconomic effects of fiscal incentives to promote electric vehicles in Iceland: Implications for government and consumer costs. *Energy Policy, 114*, 431–443. DOI: 10.1016/j.enpol.2017.12.034

Shah, K. J., Pan, S. Y., Lee, I., Kim, H., You, Z., Zheng, J. M., & Chiang, P. C. (2021). Green transportation for sustainability: Review of current barriers, strategies, and innovative technologies. *Journal of Cleaner Production, 326*, 129392. DOI: 10.1016/j.jclepro.2021.129392

Singh, A. R., Vishnuram, P., Alagarsamy, S., Bajaj, M., Blazek, V., Damaj, I., Rathore, R. S., Al-Wesabi, F. N., & Othman, K. M. (2024). Electric vehicle charging technologies, infrastructure expansion, grid integration strategies, and their role in promoting sustainable e-mobility. *Alexandria Engineering Journal, 105*, 300–330. DOI: 10.1016/j.aej.2024.06.093

Sultana, S., Salon, D., & Kuby, M. (2019). Transportation sustainability in the urban context: A comprehensive review. *Urban Geography, 40*(3), 279–308. DOI: 10.1080/02723638.2017.1395635

Sumitkumar, R., & Al-Sumaiti, A. S. (2024). Shared autonomous electric vehicle: Towards social economy of energy and mobility from power-transportation nexus perspective. *Renewable & Sustainable Energy Reviews, 197*, 114381. DOI: 10.1016/j.rser.2024.114381

Trencher, G., & Wesseling, J. (2022). Roadblocks to fuel-cell electric vehicle diffusion: Evidence from Germany, Japan and California. *Transportation Research Part D, Transport and Environment, 112*, 103458. DOI: 10.1016/j.trd.2022.103458

Tu, J. C., & Yang, C. (2019). Key factors influencing consumers' purchase of electric vehicles. *Sustainability (Basel), 11*(14), 3863. DOI: 10.3390/su11143863

Tundys, B., & Wiśniewski, T. (2023). Smart Mobility for Smart Cities—Electromobility Solution Analysis and Development Directions. *Energies, 16*(4), 1958. DOI: 10.3390/en16041958

Un-Noor, F., Padmanaban, S., Mihet-Popa, L., Mollah, M. N., & Hossain, E. (2017). A comprehensive study of key electric vehicle (EV) components, technologies, challenges, impacts, and future direction of development. *Energies, 10*(8), 1217. DOI: 10.3390/en10081217

Varghese, A. M., Menon, N., & Ermagun, A. (2024). Equitable distribution of electric vehicle charging infrastructure: A systematic review. *Renewable & Sustainable Energy Reviews, 206*, 114825. DOI: 10.1016/j.rser.2024.114825

Visaria, A. A., Jensen, A. F., Thorhauge, M., & Mabit, S. E. (2022). User preferences for EV charging, pricing schemes, and charging infrastructure. *Transportation Research Part A, Policy and Practice, 165*, 120–143. DOI: 10.1016/j.tra.2022.08.013

Wüstenhagen, S., & Kirschstein, T. (2024). Substitution of Conventional Vehicles in Municipal Mobility. *Sustainability (Basel), 16*(14), 6054. DOI: 10.3390/su16146054

Yu, J. J., Tang, C. S., Li, M. K., & Shen, Z. J. M. (2022). Coordinating installation of electric vehicle charging stations between governments and automakers. *Production and Operations Management, 31*(2), 681–696. DOI: 10.1111/poms.13564

Zentani, A., Almaktoof, A., & Kahn, M. T. (2024). A Comprehensive Review of Developments in Electric Vehicles Fast Charging Technology. *Applied Sciences (Basel, Switzerland), 14*(11), 4728. DOI: 10.3390/app14114728

Zhang, C., Yan, J., & You, F. (2023). Critical metal requirement for clean energy transition: A quantitative review on the case of transportation electrification. *Advances in Applied Energy, 9*, 100116. DOI: 10.1016/j.adapen.2022.100116

Zhao, X., Li, X., Jiao, D., Mao, Y., Sun, J., & Liu, G. (2024). Policy incentives and electric vehicle adoption in China: From a perspective of policy mixes. *Transportation Research Part A, Policy and Practice, 190*, 104235. DOI: 10.1016/j.tra.2024.104235

Chapter 3
Sustainability and Environmental Impact of Urban Mobility:
Driving Green Transportation for the Future

V. Padmavathi
https://orcid.org/0000-0001-5299-8472
A.V.C. College of Engineering, India

S. Saravanan Tamilnadu
Annamalai University, India

P. Anbalagan Tamilnadu
Annamalai University, India

B. Balasaigayathri
Sathyabama Institute of Science and Technology, India

ABSTRACT

The availability of new opportunities and the constant growth of the megalopolis challenge advertisers' sustainable travel based on transport, the volume of which, according to experts, has become one of the main factors affecting greenhouse gas emissions, air pollution, and energy consumption. This chapter aims to cover a crucial area of research that focuses on city mobility concerning achieving and managing sustainability and its impacts on the environment. It covers the trend towards environmentally friendly means of transport, including EVs, hydrogen vehicles, and using green power in city transport. This chapter also addresses ITS concerns

DOI: 10.4018/979-8-3373-0882-1.ch003

regarding traffic management, congestion, and energy use reduction efficiencies. Describing schemes such as smart infrastructure for green transport, energy-effective design, and zero carbon transport policies, this chapter offers perceptive scenarios on how cities may facilitate the transformation to sustainable low-emission transport systems with a vision for the global solution to climate change and improved environmental sustainability.

1. INTRODUCTION

Over the decades, urban structures similarly apply pressure on the need to establish competent and effective modes of transport. The environmental, economic, and social costs of traditional modes of transport have continued to grow and impose pressure on many cities globally through factors such as traffic congestion, air pollution, and greenhouse gas emissions (Fan et al., 2024). Such challenges empathically influence the quality of living in urban areas, the supply of fresh air, times of climate change, global warming, and resource depletion. Ensuring sustainable urban mobility is about identifying environmentally friendly transportation systems that will provide adequate and, more importantly, economical and easily accessible means of transport to the users within the city. This section will explain the rationale for moving towards sustainable urban mobility, stressing improving cities' sustainability, properly understood as the long-term and cumulative ability of individuals, systems, and the environment to support and meet people's needs, along with emissions reduction and upgraded quality of life.

1.1 Environmental Challenges in Urban Transportation

Transportation systems, particularly road transport, are one of the leading human-influenced causes of environmental deterioration, contributing approximately 23% of the total energy-related carbon dioxide emissions globally. High use of internal combustion engine vehicles causes air pollution, and quantities such as NO_2, SO_2, and CO_2 affect humans and the environment (Borchers et al., 2024). Secondly, road congestion is rife, leading to more fuel use, gas emissions, and waste resources. Environmental pollution also results from extended urban development coupled with small, undeveloped public means of transport. In this section, the following environmental issues specific to urban transportation will be expounded: Air quality deterioration, resource utilization, and the overall effects of climate change.

1.2 Defining Green Transportation and Its Benefits

Environmentally friendly transport, recommended as sustainable or low-carbon transport, has less impact on the environment than any other transport system. It incorporates the application of electric and hydrogen types of cars, funding the development of public transport, bicycle, and non-motorized transport systems, and adopting intelligent transportation systems (ITS) for smarter and greener transport(Zhu & Zhu, 2023). Green transportation also focuses on modern policies and technologies that decrease the utilization of fossil fuels and increase effectiveness, as well as carbon-fee and renewable energy-based policies and technologies. This section will clarify what green transportation means and explain the rationales that cover the emission of greenhouse gases, quality of air, health, and quality of urban structure. By doing this for green transportation, we will underline how this creates a policy and technological imperative to design the future of sustainable urban mobility.

2. THE ENVIRONMENTAL FOOTPRINT OF URBAN MOBILITY

Transportation, being a prominent sector, is responsible for the emission of nearly 24% of total carbon emissions worldwide (Perboli et al., 2022) (Budiman et al., 2024). This contribution is mainly important in urban areas due to high traffic congestion, industries, and population pressure. Data shows that around 97% of all urban transportation emissions result from using ICE vehicles that rely on fossil fuels (Giannini et al., 2023). These emissions include carbon dioxide (CO2), the main cause of the greenhouse effect. This section will examine how private cars, buses, and trucks influence city carbon emissions and determine actions toward minimizing carbon emissions using EVs, hydrogen technology, and efficient public transport. Furthermore, we will analyze the effect of traffic congestion on carbon emissions, showing how congestion increases emissions by making cars queue or run incorrectly.

2.1 Urban Air Quality and Pollution from Vehicles

A vehicle is one of the biggest uncontrolled sources of air pollutants; it releases NO_2, particulate matter, CO_2, and SO_2, particularly in urban areas. Such pollutants emit adverse health impacts such as respiratory ailments, cardiovascular disorders, and premature mortality, chiefly for residents of densely populated cities. Cars that run on diesel fuel release particulate matter, most of which originates from diesel; cars that use gasoline, on the other hand, release NO2, most of which is from gasoline cars. It is also involved in smog formation, disputes over urban heat islands, and the occurrence of acid rain(Asadi et al., 2022). This section will examine how transportation is a causative factor in air pollution, the implications of poor air quality in society and the environment, and what zero-emission vehicles, green transport measures, and enhanced fuel quality are in tackling air pollution in cities.

2.2 Energy Consumption in Urban Transport Systems

Most urban transportation is fuel-intensive and mainly depends on fossil fuels like petrol and diesel, which are inelastic environmental inputs and deplete resources. The author also pointed out that as cities grow, transportation energy requirements are on the rise, constantly causing the depletion of natural resources. Transportation networks that are poorly designed and rely heavily on the private car for mobility are characterized by high energy consumption levels and, therefore, are responsible for high emissions rates and high levels of reliance on fossil fuels. On the other hand, public transport, for example, is designed with energy efficiency, bringing the percent energy consumption and oil dependency down. In this section, the author will discuss how energy consumption is linked to urban mobility and how structures such as battery-powered buses, subways, and cycling infrastructure can reduce energy consumption. We will also explain how incorporating renewable energy systems into transport facilities-regardless of transport- like solar electricity charging areas for electric cars, reduces the negative effect associated with urban transport.

2.3 Noise Pollution and Other Externalities

While most people may agree that exhaust fumes are a problem regarding the environmental impact of automobile transport, it is rarely mentioned that noise is another important parameter that greatly influences people's health in cities. The steady noise road transport systems produce impacts stress levels, sleep pattern interrupted interferences, and even hearing impairment and increased propensity of cardiovascular diseases (Hwang et al., 2022). The main sources of noise pollution for urban transport include carcinogens from road traffic, such as cars, trucks, and motorcycles, as well as from rail and air transport. Hence, noise is compounded, particularly in areas inhabited by a large population and where vehicles are on the move most of the time. Aside from the noise, transport systems in urban areas also have other social costs, such as changes in land uses, loss of biodiversity, and degradation of green areas (Kumar et al., 2022). This section will discuss other negative effects of noise pollution and other social costs of urban mobility. It will also outline ways of controlling noise, including using low-noise pavements, encouraging the use of electric vehicles, and proper urban layout and design to reduce impacts from movement.

3. ELECTRIFICATION OF URBAN MOBILITY

Bio and electric vehicles, or EVs, have emerged as one of the most significant transformations to cut down the negative impacts of traffic in a city. As battery technology improves, the costs are reduced, and the demand for environmentally friendly transport rises, EVs are on the cusp of changing the outlook of city transportation (Al-Rahamneh et al., 2022). The expansion of electric vehicles depends on a regulating change that can be precipitated by advances in electric-natured drive trains, fiscal incentives, and more elevated emulation standards, as well as customer push for ecologically benign automobiles instead of fossil-powered vehicles. This section will identify some drivers that have led to the adoption of EVs, ranging from the market demand and consumers' preference to the changing nature of manufacturers in transitioning to electric vehicles (Kang et al., 2024). Furthermore, it will assess the role of emerging global trends consequent to the decarbonization of transport, notably the popularization of EVs in the urban environment.

3.1 Environmental Benefits of EVs

Some of the benefits of electric vehicles that accrue beyond the abilities of these automobiles to decrease greenhouse gas emissions in cities are as follows. Unlike ICE cars, EVs have no emissions control and thus do not emit pollutants such as NO2 and PM, negatively affecting air quality in urban settings. Moving to electric cars saves the air quality in cities and families' health expenses (Sengar et al., 2024). Moreover, developing countries can compete to produce cheap energy sources to enhance EV utilization since it may reduce global carbon emissions. This section will show that the use of electric vehicles, as well as the adoption of electric cars, can lower air pollution in urban areas, greenhouse gas emissions, and noise pollution. We will also briefly explore how EVs can help create a cleaner energy grid demand, especially as renewable energy sources are gradually integrated into the urban landscape.

3.2 Charging Infrastructure and Grid Integration

One of the key enablers of the electrification of urban mobility is the development of robust charging infrastructure. A comprehensive network of public and private charging stations is crucial to supporting the widespread adoption of EVs. This includes fast charging stations, home charging solutions, and workplace charging (Guo et al., 2023). Efficient integration of charging stations with the electricity grid is also essential to ensure that demand for electric vehicle charging can be met without overburdening the existing infrastructure. Additionally, smart grid technologies allow for better coordination between EV charging and renewable energy production, enhancing the sustainability of the entire transportation system. This section will explore the current state of charging infrastructure, the challenges in expanding it, and the role of grid integration in ensuring that electric mobility becomes a seamless part of urban transport systems. We will also discuss emerging solutions like vehicle-to-grid (V2G) technology, where EVs can act as mobile energy storage devices, helping to stabilize the grid and store excess renewable energy.

3.3 Challenges in EV Adoption and Battery Lifecycle Management

This chapter has also identified that if there is significant progress in creating electric cars in urban regions, there must be reliable systems for charging the cars. Subsequently, implementing both fixed-place public and private charging stations and mobile solutions becomes an important question for allowing a large-scale transition to electric vehicles (Borgosano et al., 2024). These are charging places for users, home charging technology, including wireless, and charging at work. Charging stations also need to be integrated with the electricity grid so that demand for the charging of electric vehicles should never be a problem as it would put a strain on the existing electricity grid (Zang et al., 2022). Thus, smart grid technologies provide an opportunity for efficient charging of e-vehicles and efficient use of renewable energy generation to help optimize the sustainable transport sector. This section of the chapter will critically analyze the existing charging infrastructure, the obstacles to future infrastructure expansion, and the role of integrating the grid to facilitate electric mobility in urban transport systems. We will also preview such trends as vehicles-to-grid (V2G) technology, which allows EVs to function as movable energy storage systems to support the grid and store renewable energy.

3.4 Case Studies: EV Success in Major Cities

Despite these innovative advantages, electric vehicle usage has undesirable implications. They also pointed out the high initial cost, the relatively short range of most electric cars, and the issue of access to charging infrastructure as the key factors discouraging people from making a switch. Moreover, the production of EV batteries also comes with the cost of environmental impact; the extraction of materials like lithium, cobalt, or nickel and the processing required to create these batteries are resource-intensive(Ren et al., 2022). Another continuous problem is the recycling and disposal of EV batteries at the end of their lifespan, alongside the growing need for raw materials and an ever-rising number of batteries that can no longer be utilized. This section will present the key drivers of resistance to using

EVs concerning financial, technology, and infrastructure constraints (Santaella et al., 2022). It will also discuss the issues of battery production, usage, and the pollution they cause,/cycle, second-life application of EV batteries, and greener sources of metals for battery manufacturing.

Figure 1. Key Components of Electrification of Urban Mobility

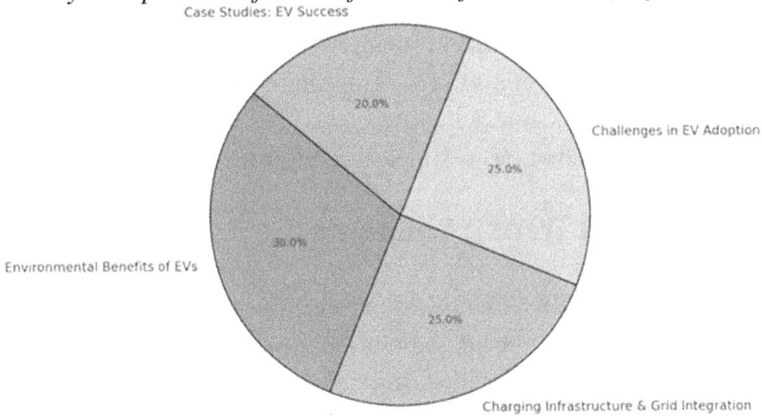

Figure 1 is a simple pie chart representing the key components of the electrification of urban mobility:

- **Environmental Benefits of EVs**: 30%
- **Charging Infrastructure & Grid Integration**: 25%
- **Challenges in EV Adoption**: 25%
- **Case Studies: EV Success**: 20%

4. PUBLIC TRANSIT: THE BACKBONE OF SUSTAINABLE MOBILITY

This chapter examines how public transport is essential in promoting sustainable urban mobility since it replaces personal car usage and reduces traffic congestion and emissions associated with personal transport (H. M. Wang et al., 2024). This section critically discusses the various ways through which public transportation contributes to the sustainability of our environment, as well as the main developments and issues.

4.1 Role of Public Transportation in Reducing Emissions

Emission Reduction Potential: Efficient public transport significantly reduces greenhouse gas emissions when incorporated into population (Nguyen et al., 2021). This is because it discourages the use of private cars in the transport system, thus reducing the number of vehicles on the road, consequently reducing traffic jams and the resulting emissions of greenhouse gases.

Urban Densification and Public Transit: When cities expand, they need well-developed public transport systems to encourage high transit-oriented development. Organizational systems promote integrated activities and densities, discouraging the formation of segregated or large, single-use buildings leading to long car travel.

4.2 Energy-Efficient Transit Solutions

Hybrid and Electric Bus Fleets: Many cities are gradually switching from conventional diesel buses to hybrid and electric buses with fewer CO2 emissions and more energy efficiency (Abdelhakim et al., 2023). Electric buses couple the zero-emissions technology at the tailpipe and can be connected with renewable energy to make public transport cleaner.

Optimized Routes and Scheduling: Measures like real-time data acquisition and predictive analysis mean that transit authorities can better position the buses, avoid periods of inactivity, and generally run more efficiently, which should translate to better energy efficiency.

4.3 Bus Rapid Transit (BRT)

BRT strategies must create high-capacity, high-efficiency public transport infrastructure on road corridors (Pozo et al., 2022). These structures are cheaper and more adaptable than standard rail structures and may be optimal for evolving cities, particularly those in the developing world (Chen et al., 2024). BRT systems have low overall emissions for the following reasons: congestion reduction, fuel conservation, and improvement in transit frequencies.

Benefits: There are fewer system construction requirements than light rail or metro systems, less impact on traffic congestion, and higher frequencies and reliability.

Global Examples: For instance, Curitiba in Brazil, Bogotá in Colombia, and Jakarta in Indonesia have fine perceptible and measurable reductions in travel times, traffic jams, and carbon footprints for having a BRT system.

Light Rail: Light rail is a middle ground of what is provided by bus systems and metro/subway systems. Clean energy for light rail and dedicated tracks means a friendly capacity to carry many passengers simultaneously, fighting emissions and providing fast and punctual connections in urban areas.

4.4 Electrifying Public Transit: Trends and Challenges

The Push Toward Electrification: With the increasing global concern for environmental concerns such as climate change, global urban centers are shifting to electrify our public transport sector (Ning et al., 2024). Many cities already had targets for the full electrification of the public transport system by the mid-century, and many new systems emerged, including electric buses, trolley buses, and even electric trains.

Technological Advancements: This is why younger battery technology and charging infrastructure have linked the potential of the complete electrification of public transport. From batteries to storing energy, wireless charging and fast charging stations ease decentralization of the convection to electric form, making this useful to an electric city's fleet. Electrification of fleets entails high capital investment in charging stations, depots or garages, and the power infrastructure due to power consumption (Song et al., 2024). Though electric buses offer enormous savings on emissions, they raise eyebrows regarding operation area and battery turnover within complex city transit systems. Although the cost of powering and maintaining these buses is low after the initial investment, the initial costs of purchasing electric buses and setting up appropriate infrastructure are still a substantial problem that many cities, particularly those in the Third World, need help to meet. The emergence of charging and fast-charging units has made it relatively easy for cities to adopt electric buses.

4.5 Integrating Public Transit with Green Mobility Solutions

Multimodal Integration: For the next generation of sustainable urban mobility, public transport, electric vehicles, bicycles, and walking must be properly integrated (Tang et al., 2021). Cities must develop systematically developed means of transitioning between the various modes to reduce dependence on private automobiles.

Shared Mobility and First/Last Mile Solutions: People movers are becoming interoperable with public transit mobility through bike-sharing, car rental, and hailing (Mishra et al., 2023). This, in turn, ensures that several passengers are helped in completing their errands and mobility within the society regardless of where they live, whether near bus or rail tracks.

Interest in MaaS apps allows for the planning, booking, and paying for a multimodal transportation trip, including PT, bike sharing, and EV riding, with a single solution. This relationship strengthens the trends of non-ownership car mobility.

Policy and Urban Planning: To increase the delivery of green mobility solutions in the public transport system, urban planners should adopt the following favorable policies on sustainable mobility (Butler et al., 2020). Among them are the identification of transport capacities as an efficient ecological transport network, investment, support of subsidies for technologies, and efficient zoning of the capacities, which are aimed at investments in infrastructure.

Urban transit does not simply represent the foundation of sustainable mobility in cities; it is a major enabler of green mobility solutions (Li et al., 2023). At the same time, measures of decarbonization of cities, increase in energy efficiency, electrification, energy-efficient solutions, and connection with other green mobility solutions will be the key drivers in forming the city of the future. Transportation is a vital cause of greenhouse gas emissions, traffic congestion, and poor living conditions in urban areas. If public transportation is developed and enhanced, any city can respond to climate change anxiously.

5. PUBLIC TRANSIT: THE BACKBONE OF SUSTAINABLE MOBILITY

Maintaining a good, healthy, and efficient skeletal structure is crucial to holding the body together, and consistent with our previous statements, this physical core is the backbone of sustainable mobility.

Investment in improving the public transport system is purposely made to shift to a sustainable mode of mobility in urban centers by dismantling the use of solitary automobiles, congested traffic, and emissions (Yan et al., 2021). This section examines the extent and manner in which public transportation enhances environmental sustainability, with special emphasis on innovation and challenges.

5.1 Public transportation and it is contribution to fighting emissions.

Emission Reduction Potential: The results also indicate that when PHS is integrated into urban mobility networks, CO2 emissions can be significantly reduced. Since they encourage the non-use of personal automobiles, their usage lowers traffic density and greenhouse gas emissions.

Urban Densification and Public Transit: As cities advance, they require sophisticated transport systems for nascent and intensified city layouts. All modes of transportation encourage high-density and integrated land use and, therefore, restrict mobility and car usage.

5.2 Energy-Efficient Transit Solutions

Hybrid and Electric Bus Fleets: The vast majority of big cities are moving from conventional diesel buses to hybrid and all-electric buses that are less polluting and more efficient in energy consumption (Pilati et al., 2020). Electric buses, for instance, have zero tailpipe emissions, and when the buses are charged from renewable energy sources, public transportation becomes even cleaner.

Optimized Routes and Scheduling: Modern technology's application in collecting real-time data and analyzing it as a tool for predicting the next possible event also helps transit authorities minimize bus idle times, thus increasing their overall energy usage.

Renewable Energy Integration: For instance, integrating solar or wind sources into public transport, such as using solar panels to power bus depots, train stations, or other parts of the transport system, strengthens the sustainability of public transport.

5.3 Bus Rapid Transit (BRT) Systems and Light Rail

Bus Rapid Transit: They are fixed-route transit systems whose aim and objective is to provide efficient and large-capacity passenger transport largely through segregated corridors. It is flexible and cheaper than conventional railroad infrastructure, mainly for emerging cities (S. Wang et al., 2022). BRT systems have the potential to serve fewer passengers, which leads to several negative effects, such as congestion and wastage of fuel, in addition to encouraging good and efficient transit systems.

Benefits: It is less expensive to build than a light rail or metro system, provides less road congestion, and provides more frequent and reliable services.

Light Rail: As with any transit, light rail systems are in the middle between buses and metro/subway systems (Cohen et al., 2021). Light rail utilizes cleaner energy sources, and having dedicated rails for movement can transport many people in a shorter time through the cities with fewer emissions.

Environmental Benefits: ELRs can be powered by renewable energy sources, making them a low-emission mode of transit compared to other mass transit systems.

6. SYSTEM ARCHITECTURE

The architecture of a sustainable urban mobility system integrates multiple layers of technology, infrastructure, and services. The goal is to ensure smooth and efficient operation while reducing environmental impact shows in figure 2. This architecture typically includes:

- **Transportation Infrastructure**
- **Vehicle Network**
- **Intelligent Transportation Systems (ITS)**
- **Energy Supply and Distribution Network**
- **Data Analytics & Decision-Making Layer**
- **End-User Services**

Figure 2. The System Architecture

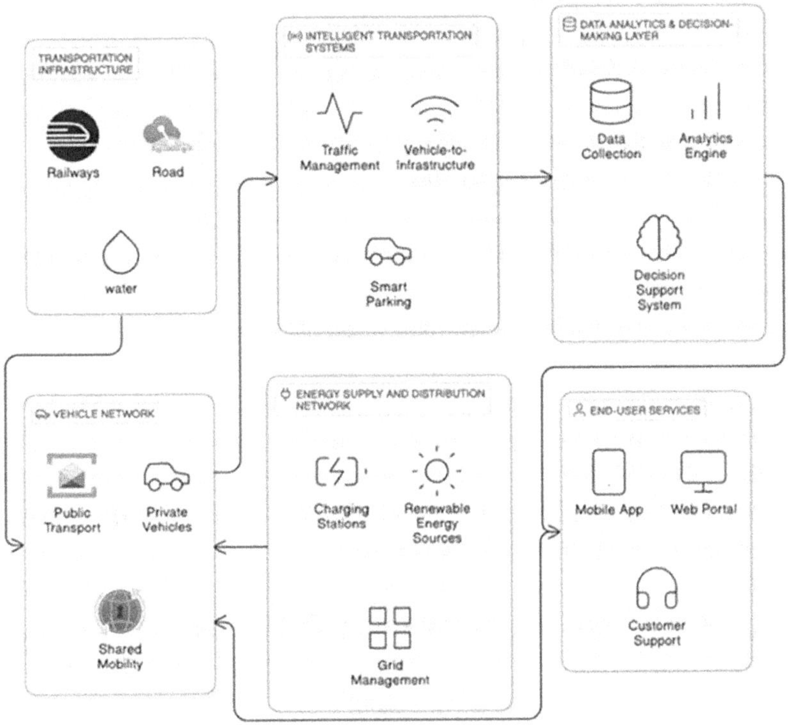

6.1. Key Components of the System Architecture

a. Transportation Infrastructure Layer

This layer involves all related physical infrastructures that facilitate the city's transport of people and materials.

- **Smart Roads:** Fitted with sensors, such as advance, traffic signals, and energy from natural resources, such as solar power.
- **Public Transit Network:** This includes buses, rail, trams, subways, and light rail with low-emitting or all-electric transport (AET) technologies. Non-Motorized Infrastructure includes bike lanes, sidewalks, and bike-sharing stands.
- **Charging Stations & Hydrogen Refueling:** Electric vehicle charging facilities and hydrogen fuel cell vehicles.

- **Smart Parking Systems:** 10 minutes of free parking and rationing of parking space for EV charging spots.

b. Vehicle Network Layer

Mobiles of people in a city or some city are represented by vehicles, which are the heart of urban mobility. This includes:

- **Electric Vehicles (EVs):** Battery electric vehicles for light, personal, shared, and public use.
- **Hydrogen-Powered Vehicles:** In the case of both private and public transport.
- **Autonomous Vehicles:** Vehicle-to-vehicle, vehicle-to-infrastructure, and vehicle-to-pedestrian communication.
- **Shared Mobility:** Transportation sharing schemes, which may include car-sharing, bike-sharing, and ride-sharing services.
- **Connected Vehicles:** Cars linked with IoT devices that, among other things, report data about traffic to infrastructure.

c. Intelligent Transportation Systems (ITS) Layer

ITS technologies allow for improving and controlling the transport flow in real time, minimizing the number of accidents and emissions.

- **Traffic Management Systems:** This includes improving traffic light phasing, easing congestion, and minimizing vehicles' time idling on roadsides.
- **Real-Time Navigation:** IoT sensors relay traffic information to car and central units for timely changes of routes where necessary.
- **Data Analytics:** Technologies include artificial intelligence and machine learning to predict traffic flow and optimize the public transport timetable.
- **Fleet Management:** Network design for routing and timetabling e-buses, shared autonomous vehicles, and other innovative means of efficient transits.

d. Energy Supply and Distribution Layer

It guarantees that the energy demand for the transport system is met optimally in the best way possible.

- **Renewable Energy Integration:** Stationary storage, solar, wind, and other renewable energy sources to complement the charging structure and the transport network system.

- **Energy Storage Systems:** Battery storage systems to store excess generated renewable energy for electrical grid balancing and to support electric vehicle charging.
- **Energy Grid Optimization:** Efficient grids that would control energy for transportation and avoid congestion in the energy grid during rush times.
- **Hydrogen Supply Chain:** Supporting structures for generating, demonization, and supplying hydrogenous mobylettres fuel.

e. Data Analytics and Decision-Making Layer

This is the intelligence in the system that forms the basis for real-time and micro-level decisions as well as macro-level strategic planning of cities and urban mobility.

- **Big Data and IoT Analytics:** Collecting, processing, and interpreting sensor, vehicle, and infrastructure readings for diagnosing utilization patterns, bottlenecks, and likely future needs.
- **AI & Machine Learning:** Divided into prediction matrix and system matrix used for traffic forecasts, optimal routing, and system redundancy.
- **Cybersecurity:** The security of the information exchanged and protection against malicious attacks, such as a hack on connected and autonomous cars.

f. End-User Services Layer

Services are available to the public and city authorities.

- **Mobile Applications:** These are applications for real-time tracking of public transport and journey sharing and emission reduction.
- **Public Awareness & Incentives:** Projects that encourage people to switch to environmentally friendly modes of transport through incentives such as tax credits or campaigns.
- **Public Transit Payment Systems:** Fare collection systems that can connect people to travel without disruptions throughout the transport medium of their preference.

6.2. Integration and Communication

The architecture of the system should be designed in such a way that all the components can communicate seamlessly:

- **IoT Connectivity:** Sensors in vehicles, infrastructure, and environmental monitoring devices will communicate with each other through IoT networks to collect data in real time.
- **Cloud Computing:** Cloud platforms process data and provide central control, coordination, and demand-response systems.
- **5G Networks:** Fast communication protocols enable low-latency communication for connected vehicles and real-time traffic management (Wu & Waterson, 2022).
- **Blockchain (for security):** It will ensure data integrity and secure transactions, especially in communicating autonomous vehicles and fleet and energy management.

6.3. Security and Privacy Considerations

Since data and connected technologies are used, urban mobility systems must include robust cybersecurity measures:

- **Data Encryption:** This will protect sensitive information from hacking, especially in the case of connected and autonomous vehicles. Vehicle Authentication ensures secure communication between vehicles, charging stations, and traffic management systems.
- **Secure IoT Networks:** Protect the communication between infrastructure elements, vehicles, and central servers from cyber threats.

6.4. Overall System Workflow

Step 1: An individual opens an application to locate the best green transportation opportunities for the current conditions (bike-sharing, EV ride-hailing, or public transport).
Step 2: Real-time traffic data, weather conditions, and energy availability are evaluated.
Step 3: The system dynamically updates the user's travel plan, optimizes his route using AI, and incorporates a mix of low-emission vehicles.
Step 4: When he arrives, he'll park in an available smart parking space, where his car will be charged for an EV.
Step 5: The central system receives data to facilitate traffic optimization and future predictions.

This architecture reflects the vision of an integrated, intelligent, and sustainable urban mobility system promoting eco-friendly transportation while guaranteeing urban mobility networks' safety, efficiency, and resilience (Gao et al., 2024). It will help cities reduce their environmental impact significantly and contribute toward global sustainability goals through ITS integration with connected technologies, renewable energy, and sustainable transport modes.

7. EXPERIMENTAL RESULTS

The experimental results developed in this chapter will allow for quantitative and qualitative analyses of sustainable urban mobility solutions. The experiments are based on simulation, case studies, and pilot implementations in various cities.

7.1 Case Study Analysis

City Examples:

- Carbon emissions from urban areas decreased by 60% because of a general transition to EVs and widespread use of green energy.
- Electrification of all public buses reduced air pollutants in urban areas by 48%.
- Integrating bicycle infrastructure decreased automobile traffic by 25% with better quality urban air.

Key Findings:

Cities with adoption of electrification and improved public transit had a persistent reduction in GHG emissions

Multimodal transport systems increase productivity and access and reduce reliance on private transport.

7. 2 Simulation Results

Using a simulated city environment

ITS-based traffic optimization:

- Traffic in real-time: reduced the congestion by 35%
- Commute average time decreased by 20%

Renewable energy integration:

- Solar-powered EV charging stations reached a 25% grid dependency reduction.
- The implementation of V2G technology offered benefits to grid stabilization.

7.3 Pilot Program Results

Hybrid and Electric Bus Fleets:

The displacement of diesel buses by electric buses reduced fleet emissions by 70% (Zhang et al., 2021). Costs were decreased by 25% as fewer fuel and maintenance costs were incurred.

Shared Mobility Programs:

The bike-sharing system improved first/last-mile connectivity by 15%, increasing public transport usage.

Smart Parking Systems:

Dynamic parking solutions reduced emissions from idle vehicles by 18%. Figure 3 shows the impact.

Figure 3. Impact of Sustainable Urban Mobility Strategies

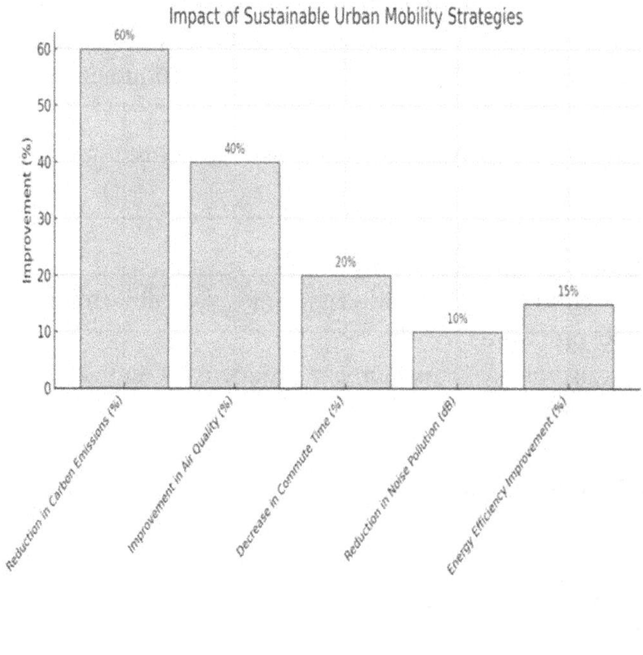

7.4 User Surveys and Public Perception

User Satisfaction:

85% of users preferred sustainable transit options for improved comfort, accessibility, and environmental benefits.

Adoption Barriers:

30% of respondents cited insufficient infrastructure as a barrier to EV adoption.

Awareness programs and subsidies significantly increased adoption rates for green transportation.

7.5 Environmental Impact Metrics

Air Quality Improvement:

Average 40% reduction in PM2.5 levels in cities with green transportation policies

Energy Efficiency:

The average is a 15% reduction in urban energy consumption from transport in systems that combine renewable energy and EVs.

Noise Pollution:

Low-noise road surfaces and the introduction of EVs decreased the average urban noise levels by 10 decibels. The experimental results strongly indicate that sustainable urban mobility strategies produce impressive environmental, economic, and social benefits. Electrification, ITS, and public transit improvements significantly reduce emissions and energy savings and improve quality of life. The findings point out a comprehensive, multi-layered approach to sustainable mobility. By scaling these solutions globally, cities can achieve climate goals and build resilient, eco-friendly urban environments.

8. CONCLUSION

Transitioning towards sustainable urban mobility is essential for alleviating all pressing challenges in urbanization, climate change, and environmental degradation. This chapter underscores how innovative transportation solutions, policy reforms, and technological advancements can greatly help mitigate the negative impact of urban mobility on the environment. Electric vehicles, hydrogen-based transport, and renewable energy integration promise to decrease greenhouse gas emissions, enhance air quality, and improve urban energy efficiency. Public transportation remains the core of sustainable mobility, backed by hybrid and electric fleets, multimodal integration, and advanced ITS that optimize traffic and reduce congestion. Furthermore, promoting non-motorized transport, such as cycling and walking, contributes to developing greener, more livable urban spaces. The experimental outcomes presented here reinforce the tangible gains from these strategies, from emission reductions and noise pollution to improved public health and quality of life. However, infrastructure investment, barriers to the adoption of EVs, and lifecycle management of their batteries require collaborative efforts involving governments, industries, and communities. The journey toward sustainable mobility is a holistic undertaking that demands coordinated action in several domains, including technology, urban planning, and public awareness. Cities that embrace these principles contribute to global environmental resilience and pave the way for a future where urban spaces are cleaner, healthier, and more equitable.

This chapter emphasizes that sustainable urban mobility is not just a solution for today's problems but also a basis for building resilient cities and fighting climate change. The closer we get to such innovations and their implementation, the closer we reach the future of green transportation and sustainable urban living.

REFERENCES

Abdelhakim, A., Soeiro, T. B., Stecca, M., & Canales, F. (2023). Multiport hybrid converter for electrified transportation systems. *IEEE Transactions on Industrial Electronics, 70*(7), 6819–6829. DOI: 10.1109/TIE.2022.3199857

Al-Rahamneh, A., Astrain, J. J., Villadangos, J., Klaina, H., Guembe, I. P., Lopez-Iturri, P., & Falcone, F. (2022). Bi2Bi communication: Toward encouragement of sustainable smart mobility. *IEEE Access : Practical Innovations, Open Solutions, 10*, 9380–9394. DOI: 10.1109/ACCESS.2022.3144643

Asadi, M., Fathy, M., Mahini, H., & Rahmani, A. M. (2022). An evolutionary game approach to safety-aware speed recommendation in fog/cloud-based intelligent transportation systems. *IEEE Transactions on Intelligent Transportation Systems, 23*(7), 7431–7440. DOI: 10.1109/TITS.2021.3069850

Borchers, T., Wittowsky, D., & Souza Fernandes, R. A. (2024). A comprehensive survey and future directions on optimising sustainable urban mobility. *IEEE Access : Practical Innovations, Open Solutions, 12*, 63023–63048. DOI: 10.1109/ACCESS.2024.3393470

Borgosano, S., Martini, D., Longo, M., & Foiadelli, F. (2024). Electrifying urban transportation: A comparative study of battery swap stations and charging infrastructure for taxis in Chicago. *IEEE Access : Practical Innovations, Open Solutions, 12*, 48017–48026. DOI: 10.1109/ACCESS.2024.3379138

Budiman, A. C., Fajratama, M. D., Ghaisani, D. N., Nurdini, A., Nurcahyo, R., & Habiburrahman, M. (2024, January). A study on electric motorcycle riding experiences and satisfaction: Perspectives from ride-hailing drivers for green urban mobility. In *2024 ASU International Conference in Emerging Technologies for Sustainability and Intelligent Systems (ICETSIS)* (pp. 1188–1191). DOI: 10.1109/ICETSIS61505.2024.10459358

Butler, L., Yigitcanlar, T., & Paz, A. (2020). Smart urban mobility innovations: A comprehensive review and evaluation. *IEEE Access : Practical Innovations, Open Solutions, 8*, 196034–196049. DOI: 10.1109/ACCESS.2020.3034596

Chen, C., Lan, Z., Zhan, G., Lyu, Y., Nie, B., & Li, S. E. (2024). Quantifying the individual differences of drivers' risk perception via potential damage risk model. *IEEE Transactions on Intelligent Transportation Systems, 25*(7), 8093–8104. DOI: 10.1109/TITS.2024.3379573

Fan, L., Wang, J., Chang, Y., Li, Y., Wang, Y., & Cao, D. (2024). 4D mmWave radar for autonomous driving perception: A comprehensive survey. *IEEE Transactions on Intelligent Vehicles*, *9*(4), 4606–4620. DOI: 10.1109/TIV.2024.3380244

Giannini, F., Franzè, G., Pupo, F., & Fortino, G. (2023). A sustainable multi-agent routing algorithm for vehicle platoons in urban networks. *IEEE Transactions on Intelligent Transportation Systems*, *24*(12), 14830–14840. DOI: 10.1109/TITS.2023.3305463

Guo, Z., Yu, K., Konstantin, K., Mumtaz, S., Wei, W., Shi, P., & Rodrigues, J. J. P. C. (2023). Deep collaborative intelligence-driven traffic forecasting in green internet of vehicles. *IEEE Transactions on Green Communications and Networking*, *7*(2), 1023–1035. DOI: 10.1109/TGCN.2022.3193849

Hwang, S., Lee, K., Jeon, H., & Kum, D. (2022). Autonomous vehicle cut-in algorithm for lane-merging scenarios via policy-based reinforcement learning nested within finite-state machine. *IEEE Transactions on Intelligent Transportation Systems*, *23*(10), 17594–17606. DOI: 10.1109/TITS.2022.3153848

Kang, J., Luo, X., Nie, J., Wu, T., Zhou, H., Wang, Y., Niyato, D., Mao, S., & Xie, S. (2024). Blockchain-based pseudonym management for vehicle twin migrations in vehicular edge metaverse. *IEEE Internet of Things Journal*, *11*(21), 34254–34269. DOI: 10.1109/JIOT.2024.3404559

Kumar, P., Kumar, R., Gupta, G. P., & Tripathi, R. (2022). BDEdge: Blockchain and deep-learning for secure edge-envisioned green CAVs. *IEEE Transactions on Green Communications and Networking*, *6*(3), 1330–1339. DOI: 10.1109/TGCN.2022.3165692

Li, Z., Liu, A., Shang, W.-L., Li, J., Lu, H., & Zhang, H. (2023). Sustainability assessment of regional transportation: An innovative fuzzy group decision-making model. *IEEE Transactions on Intelligent Transportation Systems*, *24*(12), 15959–15973. DOI: 10.1109/TITS.2023.3275141

Mishra, S., Rajendran, P. K., Vecchietti, L. F., & Har, D. (2023). Sensing accident-prone features in urban scenes for proactive driving and accident prevention. *IEEE Transactions on Intelligent Transportation Systems*, *24*(9), 9401–9414. DOI: 10.1109/TITS.2023.3271395

Nguyen, D. M., Kishk, M. A., & Alouini, M.-S. (2021). Modeling and analysis of dynamic charging for EVs: A stochastic geometry approach. *IEEE Open Journal of Vehicular Technology*, *2*, 17–44. DOI: 10.1109/OJVT.2020.3032588

Ning, X., Tian, H., Lin, Y., Yao, X., Hu, F., & Yin, Y. (2024). Research on multi-objective optimization models for intersection crossing of connected autonomous vehicles with traffic signals. *IEEE Access : Practical Innovations, Open Solutions*, *12*, 36825–36840. DOI: 10.1109/ACCESS.2024.3374041

Perboli, G., Rosano, M., & Wei, Q. (2022). A simulation-optimization approach for the management of the on-demand parcel delivery in sharing economy. *IEEE Transactions on Intelligent Transportation Systems*, *23*(8), 10570–10582. DOI: 10.1109/TITS.2021.3094851

Pilati, F., Zennaro, I., Battini, D., & Persona, A. (2020). The sustainable parcel delivery (SPD) problem: Economic and environmental considerations for 3PLs. *IEEE Access : Practical Innovations, Open Solutions*, *8*, 71880–71892. DOI: 10.1109/ACCESS.2020.2987380

Pozo, R. F., González, A. B. R., Wilby, M. R., Vinagre Díaz, J. J., & Matesanz, M. V. (2022). Prediction of on-street parking level of service based on random undersampling decision trees. *IEEE Transactions on Intelligent Transportation Systems*, *23*(7), 8327–8336. DOI: 10.1109/TITS.2021.3077985

Ren, Y., Xie, R., Yu, F. R., Huang, T., & Liu, Y. (2022). Green intelligence networking for connected and autonomous vehicles in smart cities. *IEEE Transactions on Green Communications and Networking*, *6*(3), 1591–1603. DOI: 10.1109/TGCN.2022.3148293

Santaella, J. J., Rodríguez-Bolívar, S., Puga-Pedregosa, L., González-Rico, A., Marín-González, M., & Gómez-Campos, F. M. (2022). High-luminance QD-LED device with digital and dynamic lighting functions for efficient automotive systems. *IEEE Photonics Journal*, *14*(2), 1–10. DOI: 10.1109/JPHOT.2022.3155650

Sengar, N., Kumari, I., Lee, J., & Har, D. (2024). Blockchain-based vision transformer network for safe driving perception. *IEEE Sensors Journal*, *24*(20), 32907–32918. DOI: 10.1109/JSEN.2024.3454469

Song, Z., Liu, L., Jia, F., Luo, Y., Jia, C., Zhang, G., Yang, L., & Wang, L. (2024). Robustness-aware 3D object detection in autonomous driving: A review and outlook. *IEEE Transactions on Intelligent Transportation Systems*, *25*(11), 15407–15436. DOI: 10.1109/TITS.2024.3439557

Tang, J., Wang, Y., Liu, W., Luo, X., & Zhou, Z. (2021). SCP-Tree: Finding multiple nearest parking spots with minimal group travel cost. *IEEE Access : Practical Innovations, Open Solutions*, *9*, 158946–158960. DOI: 10.1109/ACCESS.2021.3131229

Wang, H. M., Avedisov, S. S., Altintas, O., & Orosz, G. (2024). Intent sharing in cooperative maneuvering: Theory and experimental evaluation. *IEEE Transactions on Intelligent Transportation Systems, 25*(9), 12450–12463. DOI: 10.1109/TITS.2024.3379994

Wang, S., Wang, Z., Jiang, R., Yan, R., & Du, L. (2022). Trajectory jerking suppression for mixed traffic flow at a signalized intersection: A trajectory prediction based deep reinforcement learning method. *IEEE Transactions on Intelligent Transportation Systems, 23*(10), 18989–19000. DOI: 10.1109/TITS.2022.3152550

Yan, X., Allison, C. K., Fleming, J. M., Stanton, N. A., & Lot, R. (2021). The benefit of assisted and unassisted eco-driving for electrified powertrains. *IEEE Transactions on Human-Machine Systems, 51*(4), 403–407. DOI: 10.1109/THMS.2021.3086057

Zang, G., Azouigui, S., Saudrais, S., Hébert, M., & Gonçalves, W. (2022). Evaluating the understandability of light patterns and pictograms for autonomous vehicle-to-pedestrian communication functions. *IEEE Transactions on Intelligent Transportation Systems, 23*(10), 18668–18680. DOI: 10.1109/TITS.2022.3165881

Zhu, W., & Zhu, C. (2023). A prediction-based route guidance method toward intelligent and green transportation system. *IEEE Internet of Things Journal, 10*(14), 12763–12776. DOI: 10.1109/JIOT.2023.3255200

Chapter 4
Ethical and Social Implications of Urban Mobility

Maxim Kirillovich Izmaylov
https://orcid.org/0000-0002-3147-9603
Peter the Great St. Petersburg Polytechnic University, Russia

Galina Yurievna Silkina
https://orcid.org/0000-0003-0198-557X
Peter the Great St. Petersburg Polytechnic University, Russia

Maria Gennadievna Livintsova
https://orcid.org/0000-0003-3565-3880
Peter the Great St. Petersburg Polytechnic University, Russia

Viliyan Krastev
https://orcid.org/0000-0001-5456-247X
Faculty of Management, Varna, Bulgaria

ABSTRACT

Urban mobility determines the vector of economic development of territories. And if the positive aspects of it have been studied quite fully, its negative aspects have not been studied sufficiently. A systemic view of the problem of urban mobility determines not only the formulation of tasks, but also determines the methods of their solution. The chapter uses an interdisciplinary approach, including the analysis of normative documents, statistical data, a comparative study of foreign experience. Content analysis methods of interviews with experts in the field of urbanism were used to study ethical aspects. The results of the study revealed that urban mobility contributes to the deepening of social stratification in a number of cities. The

DOI: 10.4018/979-8-3373-0882-1.ch004

analysis shows that effective urban mobility management should be based on the principles of social justice, environmental sustainability, and ethical responsibility. The chapter provides recommendations for the implementation of inclusive transport solutions and emphasizes the importance of integrating ethical principles into an urban policymaking.

INTRODUCTION

Modern cities are complex social and economic systems that define trends in regional development and determine living standards. In the context of rapid urbanization and technological advances, urban mobility is becoming a key to sustainable development, affecting economic functions, as well as social structure. Despite a considerable amount of research on urban mobility positive aspects, its ethical and social impact is still understudied.

Modern cities face a whole host of challenges in transportation development, including the need to modernize infrastructure, integrate environmentally friendly transport means, ensure equal access to transport services, and minimize mobility negative environmental impacts. These processes are followed by complex social changes, specifically by growing social inequality, changes in the urban space structure, increase in digital divide, and labor relations transformation in transport and logistics.

Besides, the urban mobility concept is fundamentally changing. In addition to traditional transport mobility predicated on the principles of safety, comfort, speed and economic efficiency, new forms of mobility are becoming increasingly relevant: academic, professional, social, and virtual. Advances in digital technologies contribute to the expanding mobility concept, creating new behavior patterns and ways for people living in urban space to interact with each other. Virtual mobility operating in a barrier-free cyberphysical environment opens up new opportunities for communication and cooperation, reducing dependence on geographical location.

Aside from that, modern trends in urban redevelopment indicate a transition from traditional digitalization to intelligent systems that ensure integration of information technologies into urban space management. Intelligent mobility, based on the concept of Mobility-as-a-Service (MaaS), implies an integrated approach to transportation management aimed at increasing service availability and efficiency. However, introducing such technologies requires consideration of ethical aspects related to compliance with the social justice principles, consumer protection and ensuring transport solution inclusivity.

RESEARCH METHODS AND METHODOLOGY

The key aim of the current study is to analyze the ethical and social impact produced by urban mobility and identify global best practices that promote inclusive, sustainable and socially just transport systems.

The study methodology complies with systematic, comprehensive, and comparative analysis foundations, due to which urban mobility is comprehensively studied, including its technical, social, environmental, and economic aspects. The study is backed by the concepts of sustainable development, social justice, inclusivity, and ethical responsibility applied in modern urban studies and transportation policy.

The study reviewed international and national publications on urban mobility, working conditions in the transport sector, environmental aspects, and transport digitalization. Policy documents, such as the UNECE Handbook on Sustainable Urban Mobility and Spatial Planning, as well as country-specific legislative initiatives, were examined. Solutions used in different countries to solve ethical and social problems in urban mobility were comprehensively analyzed. Based on the data collected, conclusions on urban mobility growth prospects are drawn in terms of social justice, inclusivity, and sustainability.

RESEARCH RESULTS

Modern cities are dynamic urban ecosystems with evolving physical, social and technological infrastructure that substantially facilitates, regulates and frequently restricts free movement of residents. Urban mobility has a significant impact on the city social structure and determines the availability of economic, educational and health resources for different population groups. Urban mobility is not just about moving from one place to another; it is the possibility to access health care, education, culture, employment and leisure services using safe, environmentally friendly and affordable transport. In cities and elsewhere, mobility is one of the key tools for people to participate in public life (Sawhney, 2023).

It is important to note that transport accessibility plays a key role in urban perpetuating social inequality. Public transportation tends to develop more actively in central areas rather than in the periphery, which leads to significant differences in population mobility.

People living in cities with developed transport infrastructure have more opportunities for employment, access to educational and medical facilities, while people living in areas with a limited transport network are burdened with increased travel time, high travel costs and limited job opportunities.

Assessing public transportation quality makes it possible to determine transport accessibility in different cities and identify the efficiency of transport mobility system in terms of its convenience, reliability and urban space coverage. Russian cities with a million-plus population ranked by public transportation quality presented further reveals the overall transportation infrastructure development.

Table 1. Russian cities with a million-plus population by public transportation quality[1]

		City							Final	Position change
		Reference values		21	17	25	20	17	100	-
1	-	Moscow	13.10	19.5	15.9	22.5	19.2	12.4	89.5	-
2	-	St. Petersburg	5.60	18.8	14.6	20.0	19.4	10.6	83.5	-
3	1	Perm	1.03	15.3	15.2	18.7	19.1	9.3	77.6	Stable
4	3	Yekaterinburg	1.58	14.7	13.5	20.5	15.0	11.5	75.1	Stable
5	4	Nizhny Novgorod	1.24	16.3	16.0	15.1	16.9	7.5	71.8	↑
6	5	Kazan	1.31	14.8	11.3	13.3	17.2	9.2	70.7	↓
7	7	Volgograd	1.03	16.7	12.7	18.5	12.8	8.3	68.9	↑
8	8	Samara	1.16	15.9	11.7	18.1	13.8	8.1	67.6	Stable
9	15	Krasnoyarsk	1.20	13.9	13.5	17.5	14.5	6.8	66.2	↓
10	16	Novosibirsk	1.64	15.8	10.6	18.1	13.9	7.4	65.7	↓
11	17	Chelyabinsk	1.18	14.6	13.9	17.7	12.2	7.3	65.7	↓
12	28	Omsk	1.11	13.6	11.9	14.9	13.9	4.8	59.0	↓
13	33	Krasnodar	1.23	12.0	6.5	18.6	11.8	9.4	58.4	↓
14	34	Rostov-on-Don	1.14	12.2	8.1	16.1	15.4	6.7	58.4	↓
15	35	Tyumen	0.86	13.2	6.5	13.5	15.1	9.1	57.4	↓
16	55	Ufa	1.18	12.1	8.7	16.5	11.6	4.9	53.8	↓
17	68	Saratov	0.94	12.3	6.3	15.6	10.9	5.8	51.0	↓
18	78	Voronezh	1.05	11.8	6.5	13.5	12.3	4.0	48.0	↓

Cities with highest ranking positions (for example, Perm, Yekaterinburg, Nizhny Novgorod) show relatively high values by specific criteria, such as transportation network reliability and performance and comfort, but lag behind in other parameters, such as sustainability and safety. Thus, Perm has a high comfort value (19.1), but low value in sustainable development (9.3). This indicates that even liveable cities have lop-sided transportation development and require an integrated approach.

Transport affordability is a bottleneck for most cities. For example, for Krasnodar and Tyumen this rate is 6.5, and for Saratov – 6.3 only. This indicates that public transport fares remain high for a large part of the population, which may limit access to mobility for low-income groups. In addition, low indicators for sustainable development and safety (for example, 4.8 in Omsk, 4.9 in Ufa, and only just 4.0 in

Voronezh) indicate an inadequate attention to environmental aspects and passenger safety.

Lop-sided public transport development in cities with a million-plus population contributes to a deeper social stratification. Residents of lowest-ranked cities (for example, Voronezh, Saratov) face limited transport availability, which makes it difficult for them to access jobs, education and services. This is especially true for peripheral areas where transport infrastructure is poorly developed. Peripheral areas, especially small and medium-sized cities, are much like rural areas, which is reflected in their social structure and service accessibility.

This is especially evident in the differences between the central and peripheral city areas, where access to mobility is not only a matter of convenience, but a social inequality factor as well. Until now, new neighborhood units are being built in cities of different sizes (in the periphery as a rule), and the areas adjacent to them are being used to construct private house buildings. As a result, the gap between the housebuilding and transport, engineering and social infrastructure capacity is continuously growing (Leonchik et al., 2019). Areas previously used for industrial production are free and can be transformed into equipped public spaces. Meanwhile, city centers are often overloaded due to the large number of duplicate routes. For example, a kicksharing study held by MTS Yurent in 2024 found that 50% residents of Russian cities spend 40 to 60 minutes to commute. This creates a significant gap in the availability of time that can be devoted to education, leisure or family.

Fare and public transportation availability can be meaningful barriers for low-income residents. High transport costs can limit their mobility, reducing access to economic and social opportunities. This is especially true for peripheral areas, where residents are forced to spend a significant part of their budget on transportation costs, which increases their social vulnerability.

Comparing the growth rate of real disposable personal incomes with the dynamics of price indices for passenger transportation services (Table 2), it can be noted that real personal incomes increased by 5.4% in 2023, while the total price index for transportation services amounted to 115.7% by December 2022 and 113.6% yearly average. This implies that the price growth rates for transportation services leap ahead of personal income growth rates, which in turn reduces transport mobility availability, especially for socially vulnerable groups.

Table 2. Price indices (tariffs) for passenger transportation services in 2023 in Russia[2]

Services	By December 2022	January – December 2023; by January – December 2022
Passenger transportation services	115.7	113.6
Motor transport	113.6	110.5
City bus trip, per a ride	111.2	109.5
Fixed-route taxi trip, per a ride	109.3	108.6
Taxi fare, per 1 km	121.9	115.2
Intercity-service bus trip, per 50 km	113.1	110.5
Urban electric transport	114.3	109.9
Tram trip, per a ride	113.6	109.7
Trolleybus trip, per a ride	111.4	110.6
Subway fare, per a ride	116.0	110.0
Monthly combined ticket for traveling on urban passenger transport, ticket	114.5	110.6

Analyzing the price behavior by transport means, it can be noted that the most significant fare increase was registered in the taxi sector (by 21.9% by December 2022 and 15.2% yearly average), as well as for subway fare (by 16% and 10%, respectively). This suggests that even relatively affordable transportation means, such as urban electric transport, have become less accessible to the public.

This gap is particularly critical for people living in peripheral areas with a higher transport dependence, and limited opportunities (for example, a lack of cheaper transportation alternatives). Together, these factors lead to a decrease in low-income population mobility, which restricts their access to jobs and social services and contributes to a deeper social stratification.

All these factors negatively affect public transportation reputation. Poor transportation quality, lack of accessibility, high travel time, and lack of comfort leads to perceiving public transport as a 'second-class' travel mode. This is most noticeable in lowest-ranked cities, such as Voronezh, Saratov and Ufa, with final transport quality rates not exceeding 50-55 points. In such conditions, public transport becomes less attractive to the public, which exacerbates social stratification.

Sociological research held by the Center for Social Research of the Institute of History and Sociology of Udmurt State University confirms that the use of public transport is closely related to a financial status[3]. People who assess their financial status as average are more likely to choose buses, electric buses, trolleybuses and trams for daily trips. The reason is that this social group tends not to have access to private vehicles. On the contrary, people with a higher financial status tend to use private cars or taxis, car sharing, which reveals their higher social status and independence from public transport. The most common reasons for choosing public transport are the lack of a private car, efficiency and benefits provided for travelling by public transport.

The introduction of paid services such as car sharing, taxis and MaaS models could exacerbate social inequalities. These services often target more well-off people, leaving low-income groups without access to modern and convenient transport means.

On a separate note, ethical and social impacts produced by transport environmental initiatives should be focused. Environmental initiatives in the transportation sector, such as new electric vehicle introduction, bicycle infrastructure development and the restriction in private car use in city centers, are certainly having a positive environmental impact. These initiatives contribute to reducing greenhouse gas emissions, improving air quality and reducing noise pollution in urban environments (Nurulin et al., 2020). However, such initiatives tend to ignore the interests of people with limited mobility, including the elderly, the disabled, families with young children and other groups of citizens having difficulty with mobility. This creates new barriers to their mobility and exacerbates social inequalities. It is socially important for the disabled and other people with limited mobility to have access to transport infrastructure and railway rolling stock, since such accessibility gives them equal opportunities for community engagement.

One of the key environmental initiatives is introducing pedestrian areas, paid parking spaces and restricting car traffic in the central parts of cities. Although such measures contribute to improving air quality and reducing noise pollution, they can create additional barriers to people with limited mobility if not followed with appropriate accessibility measures.

More importantly, the transition to electric vehicles, which is actively supported by Russian and many other Governments, does not always take into account the interests of the elderly and disabled as well. For example, electric vehicle charging station are often located in places inaccessible to people with limited mobility.

The transition to environmentally friendly public transport means, such as electric buses and trams, is a key part of environmental policy. However, although new vehicles are equipped with the devices required for passengers with limited mobility, have a low floor level in the front and middle parts of the vehicle, have a spacious standing area, equipped with special wheelchair fasteners, and a ramp for entry/exit, bus stop shelters are not inadequately adapted to the needs of people with limited mobility. In particular, many cities have no platforms facilitating picking up and dropping off wheelchair passengers, while curb stones at bus stops remain too high or, on the contrary, not raised enough for convenient entry into the transport means. Highly developed driving skills are also required, otherwise it will not be easy for a wheelchair user to get into the vehicle without assistance. Who should help the disabled is still an open question. A driver has to leave his cab to do this, while conductors are normally women, and they are unlikely to be able to help a wheelchair passenger.

In general, despite technological advances in environmentally friendly public transport, infrastructure barriers continue to reduce its accessibility for a significant part of the population, which significantly limits their ability to use even those vehicles that are technically adapted.

It is also worth dwelling on digitalization and its role in urban mobility growth. In recent years, digital platforms have become vital to coordinating urban mobility. Services for calling taxis, delivering food and goods (for example, Uber, Yandex. Taxi, Delivery Club) have radically changed the transportation landscape in cities, having ensured service efficiency and convenience.

Digital platforms optimize the distribution of orders through algorithms, which can significantly reduce waiting times and increase service efficiency (Riemens et al., 2021). According to various analytical agencies, the digital service rate in urban transportation has grown to 60–70% in Russian metropolitan cities over the past five years. In the meantime, the number of drivers and couriers working through these platforms increased by 30-40%, which indicates a large-scale transition of traditional carriers to digital operation modes.

However, alongside with the positive effects, there are new problems associated with the deterioration of working conditions for drivers and couriers engaged in passenger transportation and delivery. Therefore, despite technological advantages, transition to digital platforms is associated with deterioration in working conditions.

As part of the algorithmic system, relationships between people are replaced by the interaction of an individual employee with a technical system (digital platform). This applies to both the top-down system and relationships with colleagues. The algorithm is a supervisor, and employees are organizationally and spatially isolated from each other. In particular, they are deprived of physical working places (like a workshop or office), with cooperative relations, sympathy and resistance to management being built. One can talk about the management system dehumanization in its truest sense – as a systematic elimination of human interactions.

Working for a traditional company, an employee regularly communicates with managers and colleagues; they can consult with each other, get support, feel involved, challenge the decisions made, make a compromise, or just talk to each other. Within a decentralized and depersonalized digital system, drivers find themselves isolated one-on-one with all the problems they face. The situation is a bit better for those who work in real taxi companies, as they have the opportunity to communicate with real people. Drivers struggle a lack of direct personal contact and communication both professionally and emotionally. If drivers have any service issues: technical problems, payment problems, conflicts with passengers, etc., they should contact the aggregator support service. Each aggregator works in a different way, each has a support chat, but not each has a phone line available. Most frequently, only high-status individuals can request a callback. The chatbots can talk by using standard phrases only are familiar to all drivers.

According to a study of taxi drivers (exemplified by leading services), drivers mentioned situations when they needed emergency assistance, but could not get it. In addition, they assert that aggregator companies tend to resolve conflict situations to the benefit of a passenger and stick to the 'client is always right' rule (Shevchuk et al., 2021).

One of the key challenges facing the current market for digital passenger transportation platforms is setting travel fare. Taxi services cannot independently decide on the tariff policy, since legally they are not carriers, but operate as intermediaries providing a platform for driver-passenger interaction. Such a business model compares a status of aggregators to that of an 'announcement board,' and leads to a few economic and social problems affecting both drivers and transportation service users.

One of the most critical issues occurring because of taxi aggregator operation is the economic instability of drivers' earnings. The issue is determined by several interrelated factors:

1. Aggregators' price dumping. Many drivers complain of the price dumping policies pursued by the platforms, which forces them to agree on orders at a reduced cost. This leads to deterioration in working conditions and an income decrease.
2. Competitive pressures among services. Aggregators are forced to support competitive tariffs, as reducing the travel cost is a key tool to compete for a user. As a result, there is a vicious circle created: platforms reduce prices to increase demand, but struggle an outflow of drivers, which, in turn, makes service quality poorer.
3. Obligation to complete orders regardless of their costs. It is important for services to ensure high taxi accessibility for users, including the ability to have a ride even if the trip cost is low. However, such a model often ignores drivers' economic interests who are forced to complete orders that do not even cover basic car operating costs.
4. Comparison of trip costs with public transport costs. In some cases, taxi fares are comparable to public transport prices. This is beneficial for passengers, as it provides more comfort at a minimal cost, though it poses a threat to two transportation sectors. First, public transportation loses passengers, which reduces its profitability. Secondly, the decline in taxi driver income undermines passenger transportation industry stability.

In the context of high competition between aggregators and tariff policy instability, market self-regulation proves to be impossible. An increase in tariffs by one of the services immediately leads to an outflow of passengers in favor of cheaper platforms, which forces operators to keep prices extremely low. In such a situation, government intervention is needed to ensure industry sustainability, which may include the following measures:

1. Setting a minimum charge to be paid. Fixing the minimum fare applicable to all aggregators will eliminate price dumping and provide drivers with a revenue protection guarantee.
2. Establishing a minimum hourly wage for drivers. In equivalent to the best practices of some foreign countries, taxi drivers should be paid a minimum charge for each work hour, including not only the time to complete an order, but the stand-by period as well. Such a mechanism will reduce economic instability and make a driver's profession more attractive.
3. Limiting aggregators' commission. The introduction of a maximum commission fee interest that platforms can charge drivers will prevent profit redistribution in favor of services and ensure a fair wage system.

One of the central problems resulting from active digital platform development is that employees collaborating with taxi aggregators, delivery services and other online urban mobility services are deprived of social protection that formally employed citizens are provided with. Platform workers tend not to sign employment agreements and operate within the confines of the civil law as self-employed, outside contractors, or as individual entrepreneurs. Having such a legal status, freelancers are solely liable to paying taxes and contributions; they tend to be paid in cash, which leads to their unwillingness to interact with authorities. Common points of concern associated with this employee status are lack of employment security and social protection, including health insurance, accident insurance and pension contributions. In traditional labor relations, employers are required to provide workers with a certain social package, including paid leave, minimum guaranteed wages and health insurance. However, digital platform models operate by applying to outsource and self-employment principles, which grant companies relief from the need to fulfill these obligations to their order executors.

Defining themselves as technical intermediaries between a customer and a contractor, aggregators formally assume no responsibility for drivers or couriers' labor conditions. As a result, they avoid direct labor obligations, including health insurance, pension contributions and overtime pay. At their own cost and expense, employees have to pay for medical care, car or bicycle services, fuel and equipment depreciation. This is particularly acute over the economic volatility periods, when a reduction in orders can lead to a significant decrease in income, with no compensation offered by the platform.

Some platforms have taken measures to provide basic social protection for their employees, but these initiatives are still insufficient. Over the past five years, Yandex Go and Glovo have introduced free accident insurance for their drivers and couriers. In 2022, Glovo also launched The Pledge global initiative, which provides social benefits for platform workers, including parental allowances, sick leave benefits, as well as payouts to relatives in case of couriers' injury.

Furthermore, non-fixed wages make workers' income unstable as it depends on a variety of factors, including order distribution algorithms, competition among industry workers, and changes in price patterns for services. This leads to the need to work more to gain reasonable earnings, which in turn worsens working conditions, increases tiredness and can negatively affect the safety of both drivers and couriers, as well as passengers or delivery service customers.

This conclusion is confirmed by the surveys on taxi drivers, which show that about 72% of drivers note an increase in workload and the need for multi-tasking work, which frequently leads to burnouts (Shevchuk et al., 2021). Drivers' schedules often go beyond the critical standards and lead to adverse effects for both drivers and passengers. Drivers who rent cars to work through aggregators are especially vulnerable to this. Shift duration is a critical factor for them, since fixed rental costs require a high work intensity. On average, the drivers state that they cover all their basic expenses and begin making a net profit after only ten hours of continuous work. In some cases, the working day can reach up to sixteen hours, which significantly exceeds the maximum permissible labor standards. Drivers using their own cars for work are less concerned as they do not incur additional rental costs; however, they also overwork and have burnouts because of their desire to maximize earnings.

One of the most disturbing effects produced by such a working mode is a decrease in concentration and an increased risk of road accidents. There were cases when drivers fell asleep while driving with passengers in the cabin, which in a number of situations led to fatal car wrecks. In this regard, authorities and aggregators have initiated overwork monitoring. One of the technical solutions suggests using a device designed to monitor the driver's condition by assessing their pupil reaction; this determines driver's tiredness and may alert them to a danger. However, the high device cost and the impossibility to adopt it in mass has frozen the project.

Today, one of the tools to monitor drivers' operation is a centralized database created in cooperation with the Moscow City Transport Agency. This system collects encrypted information about all drivers working through aggregators and prevents them from working beyond the sixteen-hour limit. However, as noted by respondents from various aggregators, the system is still ineffective. One of the reasons is insufficient computing capacity to process large amounts of operational data. In addition, the monitoring mechanism operates in Moscow only, while there is no such centralized regulation in other regions. Outside the metropolitan area, aggregators use a simpler application blocking, but this approach is ineffective as well, since a blocked driver can switch to another platform and continue working, which means that all the attempts to mitigate the problem come to nothing.

Thus, drivers' overwork remains one of the most significant problems for digital taxi platforms. Long working hours and the lack of proper rest leads to a deterioration in drivers' physical and mental state, a decrease in their concentration and, as a result, increased road accidents. This problem affects not only industry workers, but also passengers, who increasingly may have traffic accidents in case of driver tiredness.

Aggregators' attempts to independently monitor employees by blocking applications after the working time limit is exceeded have shown their inefficiency. State intervention is required here, but it remains local in the current situation and does not cover all regions of Russia. To ensure the safety of passengers and drivers, a unified federal monitoring system is to be designed to monitor working hours for all drivers, regardless of the platform they use and the region they work in. Such a system should be supported by regulatory legal acts to oblige aggregators to transfer data on drivers' work shifts to government agencies and restrict their access to orders after the established working time limit is reached.

It should be borne in mind that one of the reasons for overworking is drivers' low income if they stick to a normalized work schedule. A significant part of workers has to take on extra shifts, since basic incomes can't cover all the basic expenses.

Another problem in question is weird passengers. Some of the reasons for this is that taxi is sometimes perceived as a form of personal services assuming high service quality, regardless of the fares. Some passengers believe that paying even the minimum taxi fare gives them the right to expect a driver to pay special attention to them, as well as expect extra services provided beyond standard package. This can result in both minor conflicts (for example, the requirement to change the route, despite the data specified in the application), and more serious conflicts followed by insults and threats.

Another factor making the problem worse is the affordability of using taxi in aggregators. Tariff reduction has statistically decreased taxi fares and made taxi services more convenient than using public transport, especially when traveling in groups. This makes the service available to the public users, including people prone to marginal behavior. Drivers often have intoxicated passengers that show aggression and have inadequate reactions. This being so, the work of taxi drivers is a challenge not only in a physical way, but psychological as well, as it requires permanent monitoring and stress resistance.

A social aspect of the problem is the public image of a taxi driver as a profession. In light of the set stereotypes, this profession is not considered to be prestigious, which is sometimes the cause of customers' disregard for drivers. Some passengers are rude as they believe that a taxi driver must do what he is paid for regardless of the circumstances. Such an attitude lays the groundwork for new conflicts in which passengers, feeling right, can deliberately provoke the driver.

Potential effect produced by urban mobility digitalization is the use of platform order and pricing algorithms, which have a significant impact on both passengers and drivers. The algorithm opacity and the lack of clear regulatory arrangements risk for discrimination against workers, market monopolization and an increase in social inequality.

When distributing orders, many factors are to be taken into account: a driver location, their rating, the number of orders completed, dynamic demand and system priority. However, algorithm decision-making is not always transparent and fair.

For example, when choosing between two cars that are at a comparable distance from the passenger and have the same tariff, priority in receiving an order can be given to a car with a higher rating, a better order history, or to a driver who tends to take more expensive orders. Such an approach can lead to algorithmic discrimination, as a result of which certain drivers receive fewer orders due to the factors they may not even know they have.

In addition, platforms may adopt hidden forms of rating segregation, where drivers with low ratings or long order waiting time gradually receive fewer offers, forcing them to work longer hours to get an adequate income. In case there are no transparent criteria, this exacerbates inequality and creates a system of hidden job monopolization, with only 'high priority' drivers getting access to the best orders.

The pricing algorithm in digital transportation platforms remains closed and inaccessible for analysis, which has severe economic effects. According to an anti-monopoly authority representative, the Federal Antimonopoly Service of Russia is heading the investigation against the Yandex Taxi service related to non-transparent tariff regulations and unreasonably high aggregator commissions. A controversial issue is the dynamic pricing mechanism, which automatically regulates the ride costs depending on demand, time of day, weather conditions and a number of other factors. On the one hand, this mechanism optimizes traffic flows, but on the other hand, it can lead to significant price swings and speculative overpricing during the peak demands.

According to the Federal Antimonopoly Service of Russia, more than 300 complaints about Yandex Taxi pricing policy and driver and passenger account blocking were registered in 2023. The regulatory institution demanded that the company disclose the system principles for blocking accounts and reduce the aggregator's commission fees, which might reach 28% of the trip cost on a case-by-case basis. However, as of 2024, the company did not fully comply with the regulator's instructions.[4]

Another manifestation of algorithm discrimination is a non-transparent system of blocking users. Drivers can lose their access to the platform, with no criteria clearly defined and appeal mechanisms being available. As the representative of the Federal Antimonopoly Service of Russia noted, the regulatory institution demanded that the aggregator establish a limiting list of grounds for blocking and a mechanism for assessing their reasonability, but these requirements have not yet been met.

The blocking issue also affects passengers, as aggregators can use automated mechanisms to identify 'unwanted customers' without any explanation. This raises the issues of consumer rights protection and principle of justice in the digital economy.

Another negative factor is the aggregators' standpoint in disputes between a driver and a passenger. Over a long period, the services were focused mainly on satisfying customer interests, considering them a key audience. As a result, aggregators came down in favor of a customer when dealing with passenger complaints against the driver, with no situation analysis being done. Drivers could be fined or temporary blocked in the system even in cases they weren't to blame.

However, the situation has begun recently changing. Large aggregators began aware of the importance to maintain a balance between the interests of passengers and drivers. Some platforms have introduced protection mechanisms for taxi drivers, including the ability to countersue, analyze DVR records, and look for driver, as well as passenger rating systems. Sometimes, services block customers who systematically violate the rules for using the service and pose a threat to drivers.

The growth of the taxi industry is inevitably associated with the need to solve these problems, since the further market growth largely depends on the platform quality and their ability to ensure safety and comfort for all parties. The taxi aggregator industry will be further determined by the willingness of companies to comply with ethical principles with respect to drivers, and government agencies will develop effective regulatory mechanisms aimed at preventing conflicts and protecting the rights of all market participants.

The delivery industry is also experiencing adverse effects from digital platform introduction. In 2023, TelecomDaily conducted a survey among companies using courier delivery (online stores, logistics companies, courier services, etc.), with 65% of respondents mentioning an increase in the number of orders with no proportional increase in rest time, which is a risk for overwork and increased stress. Additionally, the average courier wage in real terms has decreased by 15% over the past 5 years, despite an increase in the total volume of orders[5]. This indicates that, although digitalization allows serving more customers, it simultaneously contributes to labor condition deterioration due to high competition and order distribution algorithms that do not take into account employees' individual needs.

These data show that, despite the obvious benefits in improving transportation and delivery efficiency, urban mobility digitalization is followed by severe social and economic problems. The key point of concern is that algorithmic optimization often ignores workers' needs and rights, which leads to a deterioration in their work quality and living conditions.

The growth of digital platforms for urban mobility has caused a significant increase in collecting, processing, and storing personal data of users. Services like taxi aggregators, car sharing, food and goods delivery services process huge amounts of information, including geolocation, credentials, ride history and orders. Although this data is used to optimize services, its unregulated accumulation and use creates substantial ethical and legal risks associated with privacy violations and potential data leakage.

Transport aggregators and delivery services record detailed travel routes of millions of users daily, storing information about the pick-up place and destination, intermediate stops, passenger preferences, payment data and user history. This data is used not only for order distribution, but also for targeted advertising, tariff personalization, and consumer behavior expectations.

In the meantime, the lack of transparent mechanisms for monitoring data processing leads to potential breach of confidentiality. In some cases, users are not aware of what data is collected, how it is used, and whom it is transferred to.

Since the special military operation began in Russia, the number of personal data leakages has increased dramatically. Hacker attacks, inadequate database protection and weak control over platform cybersecurity have resulted in some major incidents related to the breach of users' personal information. For example, as a result of a Delivery Club data leakage in 2022, more than 2.2 million orders leaked publicly. The data leaked contained usernames, phone numbers, delivery addresses, email addresses, order items, costs, order date and time, IP addresses. In the same year, a Yandex.Food data leakage affected more than 300 thousand users. The compromised data involved names, surnames, usernames, email addresses, phone numbers, Yandex ID[6]. This information can be used by criminals for social engineering, phishing attacks and personalized fraud.

These incidents show that digital urban mobility platforms do not always provide the proper personal data protection, and its leakage can lead to cause severe effects to users.

Attempts to address ethical and social implications of urban mobility in foreign countries have developed and implemented diverse strategies aimed at creating more sustainable, inclusive and equitable urban spaces. Some are investing heavily in public transport and bicycle infrastructure, while others are experimenting with new pricing policies to lower fares or ensure more affordable and convenient electric car purchase.

As early as in 2002, member states of the ECE and WHO in the European Region developed the Transport, Health and Environment Pan-European Programme (THE PEP). THE PEP provides intersectoral and intergovernmental political arrangement that promotes mobility and transport strategies that address environmental and health concerns. Over the years, THE PEP has developed various mechanisms to support member state activities. In 2020, the UN Economic Commission for Europe (UNECE) developed the 'Handbook on Sustainable Urban Mobility and Spatial Planning. Promoting active mobility', which laid down the foundations for creating inclusive, environmentally sustainable and socially equitable transportation systems. This document is an important guideline for urban policies aimed at reducing adverse effects produced by mobility, including transportation inequality, environmental threats and, deteriorating working conditions for employees engaged in the transport industry.

In this regard, a number of European legislative initiatives: the European Green Deal (EGD) and the Sustainable and Smart Mobility Strategy (SSMS), are aimed at creating a legal framework and incentives for introducing new urban transportation concepts. One of the priorities is 'to make intercity and urban mobility more sustainable and healthy.'

A smart strategy for urban transportation and mobility provides:

- making networking and automated multimodal mobility a reality, for example by giving passengers a possibility to buy tickets for multimodal trips,
- motivating innovation and using artificial intelligence (AI) data for smarter mobility; for example, by supporting drone and unmanned aerial vehicle deployment, and further actions to create a European shared mobility data space[7].

The above-mentioned initiatives to be implemented as part of the Sustainable and Smart Mobility Strategy involve not only new digital, environmentally friendly transport technologies, but innovative business models and concepts as well. In particular, one of these new concepts covering an overall urban life and transportation mobility is a 'smart city' concept (Kovačić et al., 2022).

As defined by the European Commission, a smart city is a place where traditional networks and services become more efficient through the use of digital solutions for the benefit of residents and businesses. Smart city goes beyond using digital technologies to utilize resource more efficiently and reduce emission, which implies smarter urban transportation networks, modernized water supply, lighting and heating, environmental waste management, etc.[8]

A smart city can also be defined as a city that integrates innovations in technology, telecommunications, networks, mobility, sustainable and environmental design and creates understandable, social, high-quality, environmental opportunities for their residents (Müller-Eie & Kosmidis, 2023). A smart city concept highlights the benefits of the improved economic efficiency, energy, governance and mobility. Smart cities are aimed at using a number of innovative technologies that can actually ensure energy saving, compliance with environmental standards, convenience and safety in passenger and cargo movement (see Fig. 1).

Figure 1. Smart and sustainable mobility technologies used in European cities

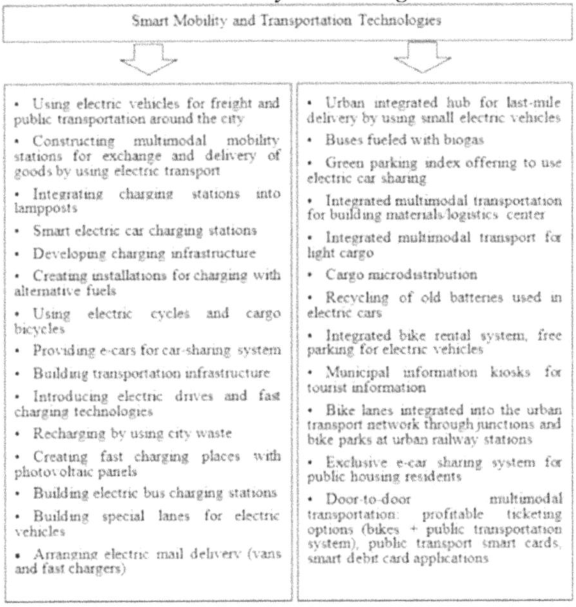

A distinctive feature of the new paradigm designed for arranging urban mobility in European cities is its fusion with spatial policies. European cities and regions transferring to mobility seek to avoid unnecessary travel and do their best to ensure various service provision near the places where people live. It can be exemplified by the 15-minute city model. The model implies that residents should spend no more time than is taken by a short walk or a bike ride from the places they live to meet their daily needs and reach infrastructure (kindergartens and schools, sports facilities, shops, medical facilities, workplaces).

This approach is successfully implemented in Paris (France), with the '15-minute city' concept being introduced (Ville du quart d'heure). The approach contributes to reducing car traffic, improving environment and life standards (Papas, 2023).

Barcelona (Spain) implements the '15-minute city' concept through the 'superblock' project (superilles). This approach aims at uniting several city blocks into one 'superblock' with the limited car movement and priority given to pedestrians and bicyclists. The outer 'superblock' boundaries remain accessible to traffic, but traffic is limited inside, thus creating safe and convenient spaces for residents. In 2016, a project was pioneered in the Poblenou region, with nine blocks having been combined into the first 'superblock' (Pupentsova et al., 2022). This pioneering made it possible to expand pedestrian zones, create new public spaces, and reduce noise and air pollution. The project success inspired local authorities to increase the initiative: 503 'superblocks' are to be created throughout the city. According to a study published in Nature Cities, the average travel time to access basic services on foot is 9 minutes in Barcelona, which is consistent with the '15-minute city' approach[9].

Milan (Italy) is also actively implementing the '15-minute city' principles as part of its sustainable growth strategy. Local authorities strive to ensure proximity of basic services and infrastructure available within 15 minutes of walking or cycling for all residents.

It is worth noting that this approach creates opportunities for solving a number of social problems: improved access for vulnerable groups to the labor market; reduced rural youth outflow, creating greater opportunities for the elderly to live longer on their own.

Urban residents need reliable, convenient and affordable MaaS founded on multimodality. Noteworthy is that alternative transport means can be successfully used only if they are followed by changes in urban resident behavior patterns. That is why local authority have the challenge to encourage residents and businesses to change their mobility patterns.

Cities and regions are implementing the MaaS concept to promote door-to-door transportation. To this end, multimodal tickets and integrated information on all possible transport means and routes are introduced.

A key sustainable mobility aspect is to ensure equal access to transport services for all social groups. This is especially important in the context of social stratification problems caused by unevenly accessible transport infrastructure. The document emphasizes the need to create affordable and convenient transport solutions for low-income groups, the disabled, the elderly, and families with children.

Practically, this is implemented in cities that introduce subsidized fares for public transport, development plans for creating affordable infrastructure and multi-level urban planning systems. For example, Stockholm local authorities (Sweden) have provided travel privileges for socially vulnerable groups, including the elderly, the disabled and people having financial struggles as part of the sustainable urban mobility concept. The system is peculiar with its provision of discounts on monthly travel tickets for people aged 65 and over. This measure contributes to reducing cost loadings on the elderly residents, in particular on people actively moving around the city, visiting medical centers, social services or meeting with relatives. That said, one-time tickets at a reduced cost for the elderly are offered only in seven regions of the country, which leaves some elderly passengers with no benefits for irregular use of public transport.

Notably, all of Sweden Public Transport Authorities that offer discounted single-ride tickets provide monthly pass discounts as well. This approach focuses on encouraging regular use of public transport among the elderly, which ultimately reduces their transport costs in the long term.

In addition, favorable travel terms apply to other vulnerable social groups. Residents who cannot work for health reasons, as well as people having serious chronic diseases that require significant financial expenses, have the same benefits as the elderly. This implies that the urban transport sector considers not only the age factor, but also the economic situation, ensuring a more inclusive and fair system of moving around the city (Bondemark et al., 2021).

Such measures contribute to increased social mobility among the elderly and the disabled. Public transport availability at reduced fares maintains their independence and helps them get engaged in public life, avoiding social isolation.

Besides, such transport policies encourage the use of public transport instead of a private car, which is in line with the urban environmental goals to reduce CO_2 emissions and improve air quality. It also reduces pressure on road infrastructure, promotes walking and cycling mobility.

In Vienna, Austria, one can buy one-time low-tariff tickets for a one-way trip, monthly or annual passes, student passes and school passes. An annual pass costs 365 Euros, that is, only 1 Euro per day. It can be bought on an installment plan that can be paid out over several months. Reduced fares are available for certain categories of passengers. Schoolchildren in Vienna can use Jugendticket (youth pass) or Top-Jugendticket (top-level youth pass).

The youth pass costs 19.60 Euros and is valid for school days from September 1 to September 15 next year. It cannot be used on weekends or during school holidays. The ticket is valid for Verkehrsverbund Ost, which covers Vienna, Lower Austria and Burgenland. This type of pass can be 'upgraded' to a senior youth pass by paying the price variations. The ticket price is 70 Euros. It is valid during the school year from September 1 to September 15 next year and can be used 24/7, including holidays, weekends and school holidays. Students under the age of 26 can buy a Semesterkarte (a semester ticket) for 75 Euros online or 78 Euros at the box office. It can be used anytime for five months from September 1 to January 31 and February 1 to June 30. During the summer holidays, students can purchase a special ticket for 29.50 Euros, which is valid for one month from July to August[10].

Tallinn (Estonia) and Aubagne (France) are prime examples of cities providing free public transport services. In 2013, Tallinn introduced a free public transportation system (Cats et al., 2017), with residents being provided with free access when receiving a personalized transit card. City visitors could also purchase a card, replenish it and use the service. The initiative was financially supported by tax subsidies with an annual budget of 12 million Euros. To ensure the system success, Tallinn enforced measures that made public transportation services more attractive. In particular, local authority introduced initiatives on additional bus stop construction, introducing real-time information at bus stops and on the Internet, as well as an increase in the number of bus lanes followed by public transportation priorities (Mateescu & Popa, 2017). These initiatives have facilitated the use of public transport by improving system convenience, reliability and efficiency (Papadakis et al., 2024).

Sustainable urban mobility plans (SUMP) recommended by the European Commission to develop all-size European cities is substantial part of the European urban mobility policy.

SUMP is a strategic plan designed to meet the mobility needs of individuals and businesses in and around cities for better life standards. It builds on the current planning approaches and considers integration, participation and evaluation principles[11].

SUMP covers the entire functional urban area with regards to real traffic flows. Urban mobility planning is based on the following principles:

- ensuring sustainable mobility throughout the 'functional urban area';
- intergovernmental synergy, including local, regional, national governments, and synergy between different policy areas;
- engaging citizens and all the parties involved;
- assessing current and future activities;
- defining long-term perspectives and clear implementation plan;
- integrated development for transport means;
- ensuring monitoring and evaluation.

Planning sustainable urban mobility has a number of key distinctions compared to traditional transportation planning (Figure 2).

Figure 2. Differences between traditional transportation planning and sustainable urban mobility planning[12]

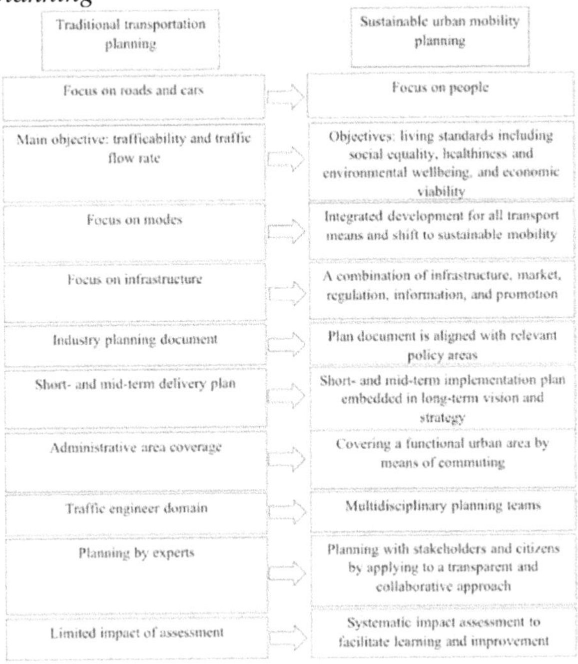

A key sustainable mobility area is developing ethical standards to manage transportation systems. This includes tariff policy transparency, protecting the rights of employees engaged in transport sector, and ensuring road safety.

Considering ethical responsibility in managing urban mobility, some foreign countries have taken measures to protect taxi drivers and couriers working through digital platforms. An example is the initiative 'Rider Law' ('Rider Law') adopted in Spain in 2021. This initiative obliged delivery and passenger transportation companies operating through digital platforms, such as Uber and Cabify, to register their drivers and couriers as employees, providing them with appropriate social benefits. The law was passed to fight against 'false self-employed', when workers are formally registered as self-employed, but actually work for hire, which deprives them of labor rights and social benefits. Under the 'Rider Law' ('Rider Law'), companies are required to provide their workers with the rights to minimum wage, paid leave, sick leave and other benefits provided by labor law[13].

In December 2024, Glovo announced their transition to an official employment model and abandoned their practice to hire couriers as 'false self-employed'[14]. This decision was made against the background of numerous fines and litigation related to labor law violation, as well as just before the company's CEO interrogation in the framework of the in criminal proceeding.

The sustainable mobility concept emphasizes the need to align transportation planning with the overall urban environment development. This indicates that transportation infrastructure is to consider population needs and integrate with residential, working and recreational areas.

Sustainable and inclusive mobility is also exemplified by Whim, a public transport, bicycle and car subscription service in Helsinki, Finland. Whim is an innovative transport subscription service launched in 2017 by Finnish company MaaS Global. It combines various transport means, including public transport, taxis, car sharing, bicycle and electric scooter rental, into a single platform with a convenient subscription system. Users can plan their trips and pay for them through the mobile application, choosing the most convenient and environmentally sustainable ways of transportation.

Depending on its type, a Whim monthly subscription could include unlimited city transport, bike hire, short-trip taxis and car hire. The platform developers expect to get people out of their personal transport by offering local residents a better alternative.

The application that changed the way of urban life is based on the MaaS concept. Whim promotes the elimination of transport inequalities by offering convenient access to different transport means for different social groups. This is particularly important for low-income residents and people having no private cars who can move around the city with no private vehicles.

Additionally, the subscription system makes transportation costs more predictable and manageable, which is especially beneficial for the elderly, students, and socially vulnerable groups. Subscription avoids violent fluctuations in travel costs, particularly significant in volatile economy periods. Although Whim operates in several European cities, Helsinki's Whim offering is the most widespread. Here, Whim offers four different models, one of which is Whim Unlimited. Whim Unlimited is a subscription package that includes public transport, city bikes, taxis, e-scooters, and rental cars on an unlimited basis with few restrictions for around 500 Euros per month.

Summarizing the above, there is a scheme compiled that reflects the ethical and social effects produced by urban mobility and solutions to be considered with regards to global best practices (see Figure 3).

Figure 3. Ethical and social effects produced by urban mobility and solutions to be considered with regards to global best practices

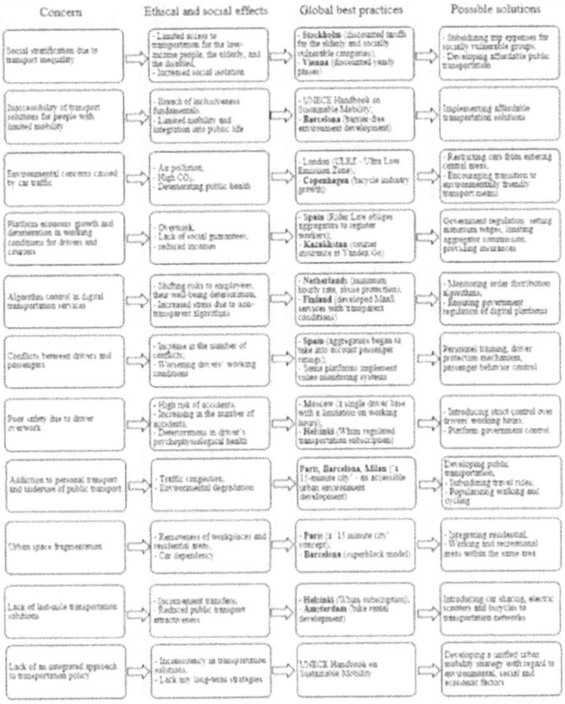

Effective urban mobility management should be rest upon the fundamentals of social justice, environmental sustainability and ethical responsibility. This assumes developing and implementing inclusive transportation solutions that consider the needs of all social groups, as well as integrating ethical principles into urban policy. In turn, ethical responsibility in managing urban mobility involves making decisions based on the principles of honesty, transparency and respect for all people's rights. When planning and administrating transportation policies, the interests of various social groups should be considered, as well as they are to be engaged in decision-making.

To summarize, a set of recommendations for implementing inclusive transportation solutions will be formulated.

1. Public engagement in planning, discussing and developing transportation projects will ensure consideration of their needs and preferences, and create more inclusive and effective urban mobility systems.

2. Developing and introducing vehicles and infrastructure that consider the needs of the disabled, the elderly, and families with children, ensures social equality principles.
3. Providing reduced fares or free travel to socially vulnerable groups; reduces financial barriers and ensures equal access to transport services.
4. Building infrastructure for bicyclists and pedestrians, as well as introducing public transport using alternative energy sources, contributes to reducing the adverse environment effects and improving public health.
5. Integrating various transport means by creating convenient transport hubs and unified payment systems can improve transport networks and facilitate movement for all social groups of residents.
6. Setting minimum wage standards for employees working throughout digital platforms and in public transport sector (including taxi drivers and couriers), regarding the order waiting time, downtime and transport contingencies.
7. Limiting taxi and delivery aggregator commissions to ensure an economically equitable income distribution between platforms and employees.
8. Ensuring strict digital control over drivers' working day duration, including mandatory rest breaks and load monitoring through state databases.
9. Providing social protection for employees employed at the platform by including them in the pension system and providing paid sick leaves.
10. Legislative consolidation of social standards established as target time costs associated with regular commuting, getting to study or getting basic services.

To ensure an integrated approach to sustainable mobility management, a new ecosystem is to enable:

- effective interaction of all parties;
- data collection in accordance with standard requirements and approved regulations;
- data analysis for each sustainable mobility goal outlined in the strategy;
- managing sustainable mobility growth.

Integrating ethical principles into urban policy requires a systematic approach and includes the following aspects:

- authorities should ensure decision-making transparency and be accountable to residents;
- when developing and implementing urban policies, the rights of all residents are to be considered and protected, ensuring non-discrimination and equal opportunities;

- urban policies should aim to improve life standards for all residents considering their needs and striving for an equitable resource and service distribution;
- decision-making should consider their environmental impact, seeking to ensure environmental sustainability.

CONCLUSION

To conclude, urban mobility is not only a matter of transportation convenience and efficiency, but also a complex social and ethical challenge. Its growth should consider the principles of equality, sustainability and inclusiveness to minimize the adverse effects produced by various social groups. The study showed that the most acute problems include social stratification when attempting to access to transport, deterioration in working conditions for workers engaged in the mobility sector, environmental threats, dependence on private transport, and insufficient integration of various transportation systems.

International experience studies have shown that best practices are being applied in different countries to address these problems. Travel subsidized programs for socially vulnerable groups in Stockholm, the '15-minute city' concept in Paris, Milan and Barcelona, the introduced 'superblock' initiative to limit car traffic, the digital platform regulation in Spain and the Netherlands – all these measures contribute to creating fair and more sustainable urban transport systems.

Particular study attention is concerned of working conditions for taxi drivers, couriers and employees engaged in the platform economy. The analysis showed that mobility digitalization leads to deterioration in employees' working condition, an increase in overwork, income instability and a lack of social benefits. In this regard, government measures regulating aggregators' activities, establishing minimum wage standards and ensuring social benefits for employees are relevant.

Thus, the study confirmed that effective urban mobility management should be based on the principles of social justice, environmental sustainability and ethical responsibility. Introducing inclusive transportation solutions, active public engagement in urban planning, protection of transport workers' rights and transition to environmentally friendly transport means should become priorities for modern cities.

REFERENCES

Bondemark, A., Andersson, H., Wretstrand, A., & Brundell-Freij, K. (2021). Is it expensive to be poor? Public transport in Sweden. *Transportation*, *48*(5), 2709–2734. DOI: 10.1007/s11116-020-10145-5

Cats, O., Susilo, Y. O., & Reimal, T. (2017). The prospects of fare-free public transport: Evidence from Tallinn. *Transportation*, *44*(5), 1083–1104. DOI: 10.1007/s11116-016-9695-5

Kovačić, M., Mutavdžija, M., & Buntak, K. (2022). New Paradigm of Sustainable Urban Mobility: Electric and Autonomous Vehicles – A Review and Bibliometric Analysis. *Sustainability (Basel)*, *14*(15), 9525. DOI: 10.3390/su14159525

Leonchik, V. P., Dolinina, O. E., and Friedrich Ebert Stiftung (2019). Sustainable mobility in livable cities. Global challenges and new approaches: [handbook on transportation system planning]. Minsk: Kolograd, 32 p. (Analysis) (Climate change, energy and environment). Bibliography in footnotes. ISBN 978-985-596-473-6.

Mateescu, C., & Popa, I. (2017). European Best Practices and Policies in Promoting Green Mobility. *EEA – Electrotehnica, Electronica. Automatica*, *65*(4), 12–16.

Müller-Eie, D., & Kosmidis, I. (2023). Sustainable mobility in smart cities: A document study of mobility initiatives of mid-sized Nordic smart cities. *European Transport Research Review*, *15*(1), 36. DOI: 10.1186/s12544-023-00610-4

Nurulin, Y. R., Skvortsova, I., & Vinogradova, E. (2020). On the issue of the green energy markets development. *Lecture Notes in Networks and Systems*, *95*, 360–367. DOI: 10.1007/978-3-030-34983-7_34

Papadakis, D. M., Savvides, A., Michael, A., & Michopoulos, A. (2024). Advancing Sustainable Urban Mobility: Insights from Best Practices and Case Studies. *Fuel Communications*, *20*, 100125. DOI: 10.1016/j.jfueco.2024.100125

Papas, Th., Basbas, S., & Campisi, T. (2023). Urban mobility evolution and the 15-minute city model: From holistic to bottom-up approach. *Transportation Research Procedia*, *69*, 544–551. DOI: 10.1016/j.trpro.2023.02.206

Pupentsova, S., Demin, A., Kirilyuk, A., & Pupentsova, V. (2022). *The introduction of digital technologies to participatory design in the public spaces formation.* In Proceedings: Innovations in Digital Economy. Proceedings of the 3rd International Scientific Conference, SPBPU IDE 2021. Ser. 'Communications in computer and information science', 325-342. DOI: DOI: 10.1007/978-3-031-14985-6_23

Riemens, R., Nast, C., Pelzer, P., & van den Hurk, M. (2021). An assessment framework for safeguarding public values on mobility platforms. *Urban Transformations*, *3*(1), 7. DOI: 10.1186/s42854-021-00023-3

Sawhney, N. (2023). Contestations in urban mobility: Rights, risks, and responsibilities for Urban AI. *AI & Society*, *38*(3), 1083–1098. DOI: 10.1007/s00146-022-01502-2

Shevchuk, A. V., Chizhova, D. A., Demina, D. E., & Chugunkova, V. A. (2021). Taxi driver operation under the algorithmic control. *Monitoring*, *3*, 356–381.

ENDNOTES

1 SIMETRA engineering company data retrieved from https://cloud.simetragroup.ru/s/jogZX2qksK3A4Ao Date of access: March 12, 2025

2 Data obtained from the Ministry of Transport of the Russian Federation. Retrieved from https://mintrans.gov.ru/file/505330 (Date of access: March 12, 2025)

3 Impact produced by public urban transport on social inequalities. Retrieved from https://docs.wixstatic.com/ugd/1cd8f0_0662585c0c6e45bfb31f08874138791f.pdf?index=true (Date of access: March 12, 2025)

4 The FAS announced serious remarks on Yandex Taxi business. Retrieved from https://iz.ru/1655571/2024-02-26/v-fas-zaiavili-o-sereznykh-zamechaniiakh-k-iandeks-taksi (Date of access: March 12, 2025)

5 TelecomDaily data analytics: 65% of companies noted an increase in courier delivery volumes over the past year, and 66% - an increase in courier costs. Retrieved from https://telecomdaily.ru/news/2023/07/18/iaa-telecomdaily-65-kompaniy-otmetili-uvelichenie-obemov-kurerskoy-dostavki-za-posledniy-god-a-66-rost-zatrat-na-kurerov (Date of access: March 12, 2025)

6 Sobetsky I. Major data leakages in Russia, in 2022. Retr-ieved from https://fbkcs.ru/publications/kommentarii-ekspertov/krupnye-utechki-dannykh-2022-goda-v-rossii/ (Date of access: March 12, 2025)

7 EC - European Commission. Mobility and Transport. Website of the European Union, 2020b. Retrieved from https://transport.ec.europa.eu/transport-themes/urban-transport_en (Date of access: March 12, 2025)

8 EC - European Commission. “Sustainable and Smart Mobility Strategy - putting European transport on track for the future”. Brussels, 9.12.2020. COM(2020) 789 final, 2020a. Retrieved from https://transport.ec.europa.eu/system/files/2021-04/2021-mobility-strategy-and-action-plan.pdf (Date of access: March 12, 2025)

9 El tiempo medio para acceder a los servicios básicos a pie en Barcelona es de 9 minutos, en Ciudad de México, 29, y en Atlanta, 50. Retrieved from https://elpais.com/clima-y-medio-ambiente/2024-09-16/el-tiempo-medio-para-acceder-a-los-servicios-basicos-a-pie-en-barcelona-es-de-9-minutos-en-ciudad-de-mexico-29-y-en-atlanta-50.html (Date of access: March 12, 2025)

10 Public Transport in Vienna. Retrieved from https://luxury-vienna.com/blog/public-transport (Date of access: March 12, 2025)

11 Sustainable urban mobility planning and monitoring. Retrieved from https://transport.ec.europa.eu/transport-themes/urban-transport/sustainable-urban-mobility-planning-and-monitoring_en (Date of access: March 12, 2025)

12 Guidelines for developing and implementing a sustainable urban mobility plan. Retrieved from https://crowplatform.com/wp-content/uploads/GUIDELINES-Developing-and-implementing-a-Sustainable-urban-mobility-plan.pdf (Date of access: March 12, 2025)

13 Ryder Law: sector light and shadows. Retrieved from https://www.abogacia.es/ru/actualidad/opinion-y-analisis/la-ley-rider-luces-y-sombras-para-el-sector/ (Date of access: March 12, 2025)

14 Globo denuncia que abandona su modelo de falsos autónomos un día antes de que su fundador declare en un proceso penal. Retrieved from https://elpais.com/economia/2024-12-02/glovo-anuncia-que-abandona-su-modelo-de-falsos-autonomos-un-dia-antes-de-que-su-fundador-declare-en-un-proceso-penal.html (Date of access: March 12, 2025)

Chapter 5
Urban Mobility Solutions as a Sociotechnical System

Yury Rifcatovich Nurulin
https://orcid.org/0000-0003-4606-1477
Institute of Mechanical Engineering, Materials, and Transport, Russia

Inga Skvortsova
Institute of Industrial Management, Economic, and Trade, Russia

Tatyana Zorina
https://orcid.org/0000-0001-9665-2756
Institute of Power Engineering of the NAS of Belarus, Belarus

ABSTRACT

The efficiency of urban mobility systems significantly depends on the following factors: what technical and technological solutions are used in the system, how the interaction of all system participants is organized and what is the main motivation of the system users. This means that when analyzing urban mobility systems, it is necessary to use a socio-technical approach. The work is based on cases of integrating sustainable transport solutions of Peter the Great St. Petersburg Polytechnic University (SPbPU) into the urban transport system of St. Petersburg. Following the socio-technical approach, the chapter presents the results of the analysis of the technical and technological subsystem of the corporate transport system with an emphasis on the infrastructure necessary to support various innovative solutions in the transport sector.

DOI: 10.4018/979-8-3373-0882-1.ch005

INTRODUCTION

The development of modern society is based on the inextricable connection of science, technology, economics and the social sphere. Urbanization has led to the fact that cities play a key role in this process. This is reflected in the concept of a “smart city”, which has become extremely popular over the past few years and is used both in scientific literature and in numerous marketing publications. This fully applies to the concept of mobility. Urban mobility is becoming increasingly “smart”, transforming itself in the direction of creating a comfortable, sustainable, greener and safer urban environment for residents. Thanks to innovations that radically change not only the properties and capabilities of individual vehicles, but also the methods of managing urban infrastructure as a whole, it becomes possible to create a sustainable and effective system of “smart” urban mobility in cities that will meet the changing needs and preferences of residents in the field of mobility. The development of innovative technologies and scientific consensus on the impact of anthropogenic factors on the degradation of the human environment, as well as the political goals of ensuring energy security and global energy competition, have generated numerous discussions on the need to accelerate and transform existing urban mobility systems (Moser, 2021; Bylinko, 2024). Modern trends in digitalization and the development of cyber-physical systems in the transport sector have led to the rapid development of concepts of “smart” urban mobility systems. It is generally recognized that the implementation of “smart” urban mobility systems can provide a number of positive results, including environmental benefits (reduced air and noise pollution in the urban environment), economic benefits (reduced costs of transporting passengers and goods, as well as repair and maintenance of vehicles, optimized use and reduced consumption of energy and fuel), social benefits (increased safety, comfort and speed of travel, increased attractiveness of urban spaces, reduced negative impact of vehicle exhaust gases on health) (Szpilko et al, 2023). The practical implementation of these opportunities requires a radical restructuring of the urban transport infrastructure. The main driving forces behind this restructuring are the following phenomena (Rodrigues, 2024).

Concentration of Population in Large Cities

Currently, more than half the world's population lives in cities, and according to World Bank forecasts, comparing to 2017, the urban population is expected to more than double by 2050 (The World Bank, 2017). The process of concentration of population in large cities leads to the growth of urban territory and the densification of its use, which requires an increase in investment in transport infrastructure, while the construction of new and maintenance of existing infrastructure becomes increasingly complex in the process of urbanization. Urban populations increase in all regions. Thus, statistical analysis of the population of large Russian cities shows a tendency of population movement to larger cities (Mkrtchyan, 2024). This causes a significant increase in the freight and passenger traffic in cities. Taking into account the population growth trend of large cities noted above, the number of private cars in cities is expected to double by 2050. Such rapid growth without corresponding changes in the urban transport system create a significant challenge for the sustainable development of cities and requests radical changes in the city infrastructure, traffic management and transportation modes (Uribe-Chavert & Posadas-Yagüe, 2025).

The concentration of people in urban areas directly affects their choice of travel modes, which may include walking, cycling, e-scooters or various forms of public transport.

The lack of appropriate and effective solutions in the field of urban mobility management leads to a significant increase in the number of cars on city streets, which ultimately inevitably leads to congestion and parking problems. All this causes a decrease in the efficiency of the system from an economic point of view, and also leads to an increase in exhaust emissions and environmental pollution. The rapid growth in the number of urban residents can overload existing public transport systems. Rapid growth in the number of city residents can overload existing public transport systems that are designed for fewer passengers and cannot cope with the increased flow, especially during peak hours. Due to increased traffic jams, vehicles are often delayed, resulting in economic losses and a decrease in the quality of transport services. Traffic congestion is a serious problem in modern cities, resulting in increased travel times, fuel consumption and environmental pollution. To adapt urban transport systems to increased passenger traffic, metro systems, light rail transport, dedicated lines for public transport and taxis, and intercepting parking lots at metro stations at the entrance to the city are used. The development of alternative modes of transport indirectly contributes to the reduction of the need to use personal transport, which reduces the severity of the traffic problem in cities.

All these solutions require significant capital expenditures and are limited by the availability of free space (Mirović, et al, 2024). The authorities of the European Union and other countries are taking active steps to reduce the scale of negative consequences of rapid development of urban transport (Sluiter, 2021). Existing city roads are struggling to cope with growing transport needs. The issues of analysis and management of urban traffic are the focus of research all over the world (Wang, et al, 2025; Rajkumar et al, 2025).

Green Energy Transition in Transport Sector

It is known that the transport sector is one of the significant sources of environmental pollution, and urban mobility exacerbates this problem due to the high concentration of pollution sources. Transition to electric transport in urban mobility is one of the tools for achieving decarbonization goals to combat climate change (Pawlowska & Suchanek, 2019). Green Urban Mobility is an integral part of the green transformation of transport as it introduces technological innovations that significantly reduce pollutant emissions, increase energy efficiency and promote more sustainable forms of urban transport. Traditional combustion engine vehicles are the main sources of CO^2 and other harmful emissions in cities (Vorwerg et al, 2025). Green urban mobility promotes alternatives such as electric cars, bicycles, e-scooters and public transport that produce much less pollution, thereby helping to reduce the city's carbon footprint.

It should be noted that green transition (including the urban mobility dimension) is an extremely complex process that includes economic, technological, environmental, administrative and social aspects. Economic aspects of the green transformation of urban mobility include analyzing costs and benefits related to introducing new technologies and transport solutions. The introduction of low-energy transport solutions and the reduction of harmful emissions can generate savings in the long term by reducing dependence on fossil fuels and costs related to environmental pollution and public health. However, this requires initial investment in infrastructure and public education. The technological aspects of the green transformation of urban mobility focus on developing and implementing modern solutions, such as electric and autonomous vehicles, intelligent traffic management systems, and digital platforms supporting vehicle sharing. The use of advanced technologies allows for the optimization of transport systems, improved safety and increased efficiency. Environmental aspects related to the green transformation of urban mobility focus on reducing greenhouse gas emissions, air pollution and noise reduction in the city. Switching to electric vehicles, bicycles, or public transport powered by renewable energy can significantly reduce the negative impact of transport on the natural environment and the health of residents. Administrative aspects of the green transformation of urban mobility include the development of appropriate policies and regulations to promote sustainable transport and support investments in transport infrastructure. The introduction of financial incentives, such as tax breaks or subsidies for the purchase of electric vehicles, and the development of urban plans favoring pedestrians, cyclists and public transport are crucial to the success of the transformation.

At the same time, for an objective assessment of the impact of certain innovative solutions on the environment, including green energy, green mobility, etc., it is necessary to systematically analyze the entire life cycle of a product or service, including production and disposal, and not just the stage of its use. The same statement applies to the economic assessment of innovative solutions: a comprehensive assessment is required, which includes not only an analysis of current tariffs and costs, but also capital costs, as well as the role and need for government subsidies and other forms of financial support.

The widespread use of personal electric vehicles requires the development of charging infrastructure and the provision of sufficient power at charging points, which may require a redesign of the urban energy system (Nurulin, Y. et al, 2021). The variety of types of infrastructure elements for charging electric vehicles, buses, trucks and e-scooters adds additional complexity to the problem of providing electrical power at urban charging points (Sankhwar, 2024). The analysis shows a significant reduction in the need for public charging stations in urban areas as households increasingly use home charging stations. This confirms the importance of the challenge of balancing residential, office and commercial charging infrastructure to meet the growing needs of electric vehicles (EV) users. The need to address this challenge forms the basis for strategic decisions by policymakers and utility providers in urban planning that support the spread of EVs and promote environmental sustainability (Tarafdar et al, 2024; Graf & Kantel, 2025).

Growing Popularity of Micromobility

Recent studies show that urban mobility patterns have changed significantly since the COVID-19 pandemic (Anke et al, 2021; Correa, et al, 2020). The use of public transport and taxis has decreased in favor of individual means of transport, which are considered safer and more hygienic (private car, bicycle, walking). Many companies have switched to remote work for employees. Overall, this has led to less regular use of public transport for commuting compared to pre-pandemic levels. Since the end of the pandemic, the popularity of active mobility, mainly cycling, walking and micromobility, has continued to grow, and the rate of this growth has increased compared to pre-pandemic levels. At the same time, the number of residents of large cities who use daily delivery of goods from supermarkets and restaurants has increased. This, in turn, has led to a sharp increase in the number of delivery drivers on city streets using micromobility vehicles (electric scooters and e-bikes).

All this creates additional pressure on the urban transport system and creates new challenges (Heumann et al, 2025).

The phenomenon of the growing popularity of micromobility is based on the widespread use of small, lightweight vehicles, powered by humans or electric motors, for individual trips at low speeds in urban areas (Dediu, 2019). Micromobility is most common in densely populated urban areas, where it helps reduce traffic and air pollution in the city (Tuli et al, 2021). By developing delivery services, it has a significant impact on the economy as a whole: business processes in trade are changing, the required retail space for stores is reduced, and turnover is accelerated.

Over the past few years, electric scooters have become the most popular micromobility solution in Russia. As experience has shown, electric scooters create both additional opportunities and additional problems for public transport systems. First of all, these problems are related to the safety of both the users of electric scooters and pedestrians, who are threatened by electric scooters. A number of countries, including Russia, are introducing administrative restrictions or direct bans at the national or local levels on the use of electric scooters in urban areas. These bans affect the maximum speed of movement, limit parking or movement in certain areas, etc. The problem of managing micromobility is especially relevant in urban areas with historical buildings, where existing streets do not accommodate new transport needs, and the lack of free space does not allow for the allocation of separate lanes for electric scooters.

Urban Mobility in the Context of Smart City Strategies

A number of studies have provided detailed analyses of smart city concepts with the focus to smart mobility issues (Albino et al 2015; Wolniak, 2023; Bylinko, 2024; Gulc et al, 2024). The concepts of smart city and urban mobility are closely related, and develop in a complementary manner, using the Internet of Things (IoT) and artificial intelligence (AI) technologies and data analysis (Gheorghe & Soica, 2025; Adewale & Paul, 2024). Among other aspects, these studies highlight the importance of the problem of understanding and supporting the very essence of a smart city by the city residents and urban communities (Mykhailova, 2023). A smart city is not only the result of a carefully planned project based on a set of innovative technologies, but also the result of the initiatives of various stakeholders with different interests (Coletta et al,2019).

Smart technical solutions for urban mobility control systems have become commonplace in modern cities: traffic light control and automatic recommendations for choosing the best route based on current traffic, road safety monitoring and informing passengers about waiting times for public transport at stops, etc. Innovative solutions such as autonomous vehicles, shared electric bicycles or cars are becoming more common in smart cities. Their effectiveness is largely determined by the effectiveness of the information technology as well as organizational and economic mechanisms for managing these systems. As a result, in the course of its development, the concept of a smart city influences the directions of development of urban mobility systems, creating a more sustainable, efficient and friendly urban environment for residents.

From the systematic point of view urban mobility in smart city is based on the following principles:

Flexibility: multiple travel options with the ability to optimize them depending on the specific situation.

Intermodality: door-to-door travel on different modes of transport with simple and convenient transfers.

Efficiency: minimizing the cost of travel to the destination in the shortest possible time and with minimal delays along the way;

Integration: technical consistency and complementarity of various urban transport means in a single urban space

Minimal impact on the environment and safety: use of clean technologies in transport with zero emissions, minimization of noise and vibrations, prevention of accidents

Smart control, which implies a high level of implementation of innovations in products, services and business processes based on ICT and AI technologies, facilitating the transition to a knowledge economy. Today, AI has become a critical factor in the development of all spheres of public life. In urban mobility management tasks, AI ensures the emergence of qualitatively new effective solutions in the field of traffic management and optimization of urban transport infrastructure. At the stage of designing an urban mobility system, the predictive capabilities of AI algorithms will help to improve the effectiveness of management decisions on the development of the urban road network, the construction of parking lots and other elements of the transport infrastructure. At the operational stage, AI uses real-time analysis of data from sensors, cameras and GPS, which allows dynamically adjusting traffic lights to optimize traffic flow, reduce congestion and create more sustainable urban transport systems. This adaptive approach to urban mobility management improves the efficiency of the transport system, reducing travel times, fuel consumption and exhaust emissions (Fatima & Rasool, 2025). AI technologies also contribute to the integration of various modes of transport into a unified urban mobility system. Joint management of different types of urban public transport, private vehicles, shared mobility and autonomous vehicles allows for the creation of a seamless and efficient urban mobility network. The use of AI in urban mobility solutions supports the development of smart cities by improving the safety of the transport sector, reducing the negative impact on the environment and increasing the efficiency of transport systems. AI-based smart traffic management systems collect and analyze data from road sensors, surveillance cameras and GPS systems to optimize vehicle flow and reduce traffic jams.

Smart city governance, which integrates various aspects of a smart city and ensures increased accessibility of public services to residents by expanding the scope of application of information systems to ensure interaction between city authorities and citizens. Smart city governance involves the participation of broad sections of the population in decision-making on the development of the city. It includes transparency of strategic decision-making procedures, accessibility and quality provision of public services. Among other things, intelligent city management develops connected public transport systems, which have significant potential to improve urban mobility management. By coordinating the work of companies providing passenger transportation by various modes of transport, city authorities can significantly influence passenger flows and the efficiency of the city's transport system as a whole. Significant attention is paid to the problem of urban parking development and parking management issues play a significant role in city governance in overcrowded cities. These systems combine organizational and economic mechanisms for managing parking in urban areas (preferences or restrictions in parking for certain categories of residents, differentiated parking costs depending on the area, day of the week or time of day, intercepting parking at metro stations at the entrance to the city, etc.). Modern smart parking solutions include information systems that contain sensors, subsystems to manage the use of parking space, mobile applications that provide parking space monitoring and booking, etc. An important aspect of effective urban mobility governance in modern conditions is the optimization of the placement of electric vehicle charging points and providing the priority of pedestrians and cyclists. Although projects to create or reconstruct sidewalks and bicycle paths require significant capital investments, they help make walking and cycling in urban conditions safer and more attractive as alternative modes of transport.

Smart economy, which is based on efficient technologies, supported by advanced ICT solutions that promote innovation in products, services and business models, facilitating the development of local and global business connections, exchange of goods, services and knowledge. Mobile applications and information systems play a significant role here, providing services to passengers, including information on routes, schedules, availability of different transport options for planning trips, purchasing tickets, sharing vehicles for both cars and e-bikes. All this ensures better accessibility and efficiency of public transport, reducing the number of vehicles on the roads. Within the general trends of urban mobility development, the popularity of shared mobility services such as bike-sharing, car-sharing and ride-hailingis are growing and can be considered as a developing type of service (Narayanan, 2023).

The results of the assessment show the impact on the efficiency of the urban mobility of socio-demographic characteristics of users (age, gender, education, availability of a car in the family, availability of a public transport pass), travel parameters (distance and travel time, comfort and safety) and accessibility (waiting time, proximity of stops or parking places). The economic efficiency of smart mobility systems largely depends on the efficiency of the mobile applications and information systems used, which provide services for passengers including information about routes, timetables, availability of various transport options for travel planning, ticket purchases, vehicle sharing for both cars and e-bikes. All this provides better accessibility and efficiency of public transport reducing the number of vehicles on the road

Smart environmental management, which is based on continuous monitoring of urban environmental pollution and energy efficiency principles for the rational use of natural resources. The use of these principles in the development of urban infrastructure involves the implementation of projects to modernize street lighting, water and electricity supply systems, as well as the renovation of buildings to reduce the resources consumed. In this series, projects on the urban transport infrastructure development occupy a special place, since urban mobility is one of the significant consumers of energy resources.

Social aspects, which are based on the principles of social diversity and creativity, that allows for the formation of individual lifestyles, behavior and consumption patterns (Ma et al, 2023). Increasing the level of accessibility of the urban transport system has a pronounced social aspect: with the growth of urbanization, problems associated with the inability or unwillingness of a certain part of the population to own or drive a vehicle are becoming more acute. However, mobility is needed by all people, and their needs must be met.

In addition to technical and organizational-economic aspects, a number of researchers analyze the socio-behavioral aspects of urban mobility (Brzeziński, 2024; Amara et al,2024). Research into the social aspects of green urban mobility allows us to take into account social diversity and the accessibility of transport for people with different needs, and to ensure that the transformation of urban mobility benefits all city residents. Understanding and taking into account typical patterns of residents' behavior is a prerequisite for making effective decisions on urban mobility management (Irvin & Stansbury, 2004). At the same time, understanding the interaction system of key stakeholders involved in the creation and development of the urban transport system is also a prerequisite for designing an effective system (Suvvari & Saxena, 2023). All of the above shows that the problem of urban mobility development goes far beyond the issues of transition of urban transport to reducing CO2 and noise emissions. It also addresses issues of social behavior and user motivation, requires active participation of city authorities and other stakeholders in making effective decisions and using modern AI solutions to change the quality of daily life of residents, which requires a comprehensive approach to the analysis and synthesis of the system of urban mobility.

The green transformation of urban mobility in the context of social aspects is a comprehensive process of adapting transport systems in cities, striving to minimize the negative impact on society and promoting sustainable development. It includes both technological and social measures aimed at reducing emissions of harmful substances, improving air quality and ensuring safe, accessible and fair access to various forms of transport for all social groups. As part of this transformation, the priority is to promote alternative means of transport, such as public transport, cycling and walking, while limiting the use of vehicles powered by fossil fuels. There are also efforts to increase the availability of transport for people with different needs, reduce road congestion and limit spatial segregation by creating urban environments conducive to social integration and improving the quality of life of residents.

The selection of social factors for further analysis is justified due to the important role of society in the process of urban mobility transformation. The urban community is the ultimate recipient of changes in the transport system, so understanding its needs, preferences and barriers is crucial for effectively planning and implementing the green transformation.

Smart urban mobility is based on a systematic consideration of the above issues. It should be noted that in a number of publications, smart urban mobility is considered as a specialized system aimed at reducing traffic jams and exhaust emissions, improving road safety, as well as increasing the efficiency and sustainability of freight and passenger transportation. Systemic studies implementing a holistic approach to the analysis and synthesis of urban mobility systems are insufficient. This work contributes to solving this problem.

METHODS

Urban mobility management systems belong to the class of complex systems. Functional complexity is determined by the inefficiency of standard strategies and approaches to solving problems of both the system as a whole and its individual subtasks. Structural complexity of urban mobility is determined by the presence of interacting subsystems of different nature in the system: technology and engineering, economics and regulatory support, the system of preferences of key stakeholders and users. The tasks that individual subsystems of urban mobility solve are not independent in preference: a solution within one subsystem can significantly affect the decisions of another. Additional complexity is introduced into the system by the very nature of the socio-behavioral subsystem and its internal properties. If the technical, technological and organizational-economic subsystems of urban mobility can be described by formal methods for which it is possible to find an optimal solution, then the behavior of users of the system is difficult to represent in the form of a formal model that adequately describes the inner world of a person, his desires, preferences and behavior. Thus, we will further consider urban mobility as a complex sociotechnical system.

The effectiveness of urban mobility systems depends significantly on the following factors: what technical and technological solutions are used in the system, how the interaction of all participants in the system is organized and what is the main motivation of the system users. This means that a sociotechnical approach must be used when analyzing urban mobility systems.

The socio-technical approach, as a doctrine of the organization of activity and a set of methods and techniques used in the tasks of analysis and synthesis of systems with special properties, has been developing since the middle of the last century. The recognized founders of research on this issue are Eric Trist and Fred Emery, who in the 1960s proposed the term "sociotechnical system" and developed a sociotechnical approach to the analysis and synthesis of organizational systems. This approach was further developed within the framework of the socio-technical model proposed by Harold Leavitt, which includes four interacting and coordinated dimensions - people, task, structure and technology - as important components of the organizational system of work (Leavitt, 1965). A number of authors present a sociotechnical system as a six-dimensional interconnected structure including goals, people, culture, technology, infrastructure and processes (Challenger & Clegg, 2013; Sony & Naik, 2020). These elements are embedded in an external environment consisting of several stakeholders, economic circumstances and a legal framework (Davis et al, 2014).

Sociotechnical systems constitute a special class of systems, the effectiveness of which equally depends on technical, organizational and social factors. Joint consideration of these factors in the analysis or synthesis of a system allows full use of the synergistic effect, which is based on a deeper understanding of the system goals, the nature of its internal structure and the interdependencies of its elements. Modern ideas about the nature of socio-technical systems are based on their description as open nonequilibrium systems, in which the imbalance can be disrupted due to a wide variety of fluctuations of both anthropogenic and technogenic nature. These systems contain groups of elements of two types. The first group consists of well-formalized elements. The basis of their activity is linear cause-and-effect relationships formed at the stage of system synthesis and predictably developed at the stage of its operation. The second group consists of weakly formalized elements. The basis of their activity is complex nonlinear relationships. The basis and framework of these relationships are laid at the stage of system synthesis, but at the stage of operation, they can develop unpredictably. Their cooperation is based on the principle of feedback, which gives the system some stability. All this additionally emphasizes the need for continuous analysis of the socio-technical system at all stages of its life cycle to make appropriate management decisions on adjusting the composition and functions implemented.

The social aspect of urban mobility reflects the behavior of people, their relationships within communities, organizations and society as a whole, the motivation system and performance indicators. Technical aspects reflect technologies, innovations, knowledge, processes and methods. The sociotechnical approach assumes that when introducing new technologies or changes in the business processes of the urban mobility system, the relevant social and technical factors should be considered in a complex at all stages of the system's life cycle, starting with pre-project investigations up to the operation of the system.

As the above analysis of literary sources shows, an intuitive understanding of the sociotechnical nature of urban mobility is present in the vast majority of studies aimed at studying the patterns of development of this system. At the same time, systemic studies of the sociotechnical nature of the urban mobility system are insufficient. Further analysis is based on the case of the implementation of a pilot project to create a corporate mobility management system in Peter the Great St. Petersburg Polytechnic University (SPbPU).

RESULTS AND DISCUSSION: CASE FROM SPbPU

Corporate mobility management system (CMMS) cares in a holistic way how employees get to work. The main goal of CMMS development is to reduce the use of individual vehicles by each employee, since public and corporate costs for mobility are high not only for the company and the city as a whole, but also for the employers themselves. On the one hand, the maintenance of each parking space on the territory costs an average of 250 euros per month per car, and on the other hand, cities need to reduce the number of individual vehicles to manage traffic jams and parking in the city. The advantage for each institution that develops its own corporate mobility management is not only that it saves own budget, but also that it ensures good working conditions and competitiveness. Good companies with a good mobility system get good employees.

The pilot project to create a corporate mobility management system at SPbPU was launched in 2021 as a further development of a holistic approach to urban management, which was the subject of analysis within the framework of the European projects AREA 21 and LUCIA implemented in 2018–2022 in the frame of European Program BSR INTERREG. The methodological basis of the project was the sociotechnical approach to the analysis and synthesis of the CMMS (fig. 1).

Figure 1. Urban mobility as a socio-technical system

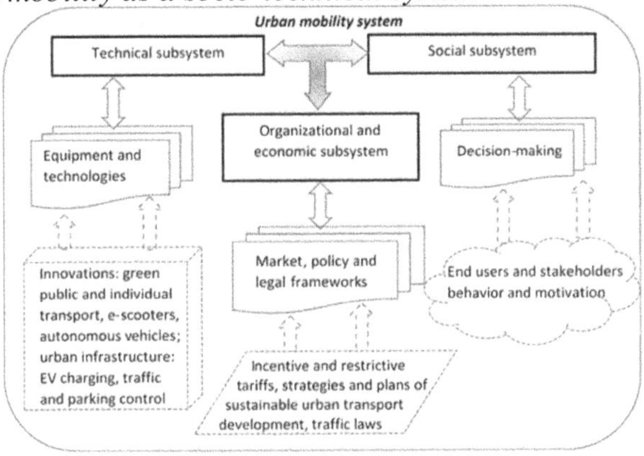

The technical subsystem of the CMMS includes passenger cars, buses, and special transport for road maintenance. This also includes the campus road network itself, prepared parking spaces, garages and service points, and a system for managing vehicle access to the campus territory. The transition to environmentally friendly modes of transport (electric vehicles or gas-powered vehicles) will require corresponding changes in the service infrastructure.

The efficiency of the corporate mobility management system depends significantly on the parameters of its external environment and the preferences of its users. From a formal point of view, the personal transport of employees is not part of the CMMS, but their number and the dynamics of using different types of transport create significant challenges for the CMMS. Therefore, the most important task at the stage of designing the corporate mobility management system is its integration into the urban transport system. An effective solution to this problem allows for a synergistic effect both for the CMMS and for the city as a whole.

The organizational and economic subsystem includes the regulatory framework governing the rules for using SPbPU transport. It determines the procedure for ordering, paying for and using transport vehicles by SPbPU Institutes and Graduate Schools, as well as forms the organizational structure of the SPbPU departments responsible for the maintenance of the CMMS. In addition, this subsystem defines the rules for access of vehicles to the campus territory. It is being developed within the framework of the current rules for the use of urban infrastructure, therefore the SPbPU departments responsible for the development of the CMMS are in constant contact with regional authorities. A significant part of the SPbPU campus is a territory included in the list of cultural and historical heritage sites at the federal level. Therefore, any work on changing any objects on this territory, including work on developing the road network, must be agreed upon and approved by the Committee for State Control, Use and Protection of Historical and Cultural Monuments. This requires studying the relevant regulatory framework governing issues of preserving cultural and historical heritage and developing instruments for cooperation between city government structures, private businesses and public organizations involved in the processes of sustainable development of urban mobility. The duration of procedures for obtaining permits from this Committee may take several months, which should be taken into account when planning the work. Additional problems may arise due to the fact, that a number of works, at the request of the Committee, must be completed only in the spring, which imposes serious restrictions on the calendar plan of works.

All decisions regarding the organization of entry and exit to the campus territory must be agreed with the city authorities (the Committee on Transport and the Committee for the Development of Transport Infrastructure of St. Petersburg). Taking into account the above, the relevant Committees of the Government of St. Petersburg must be included in the key stakeholders of the CMMS.

The social subsystem determines motivation and behavior of the CMMS users. Its task is to engage, involve, inform and listen users (students, teachers and employees of SPbPU) to the work of development and sustainable use of the CMMS.

Engage: focus on creating an attractive urban environment on SPbPU campus for students, staff and citizens, taking into account the diversity of their preference systems; increasing the efficiency and sustainability of CMMS; reducing the negative impact of CMMS on the environment.

Involve: target groups - decision makers in the city and SPbPU, experts and specialists in the development and maintenance of CMMS, end users, including students, master's and postgraduate students of various institutes, existing student communities, employees and visitors of SPbPU, residents of adjacent urban areas.

Listen: the aim is to receive feedback on the CMMS development by means of questioning of students and employees, organizing targeted seminars and round tables, discussions within the framework of student research

CITE DESCRIPTION

The pilot site for creation of the corporate mobility management system of SPbPU was the central campus (Figure 2). To one degree or another, this affects all elements of the complex socio-technical system of urban territory management, which includes mobility infrastructure.

Figure 2. SPbPU campus site

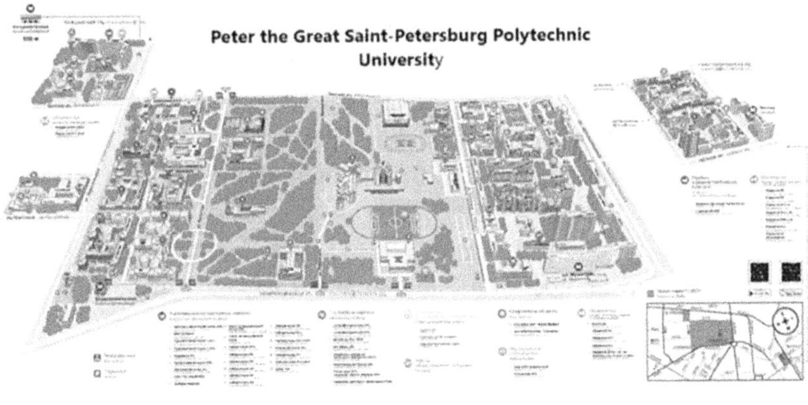

From a management perspective, the SPbPU campus is an area of over 120 hectares with over 35,000 users (students, teachers, researchers, service personnel and residents of several apartment buildings) which create a significant flow of cars and pedestrians, especially during peak hours (before and after classes).

Buildings and Transport Infrastructure

There are 112 buildings on the campus, which are located in 5 sites (see fig. 2). Half of them are pubic buildings, including 42% of educational buildings and 8% of social buildings (kindergartens, clinic and sports complex). 35% of the buildings belong to the residential area. 27% of them are dormitories and 8% - residential buildings. 15% of buildings are manufacturing and offices. Eight buildings were built 120 years ago when SPbPU was established and has a status of historic buildings of the federal level.

SPbPU is carrying out a number of projects for the construction of new and reconstruction of existing buildings on campus. Most of these projects include transport infrastructure issues.

Two metro stations are located in close proximity to the campus. Near one of them, there is an intercepting parking for 60 cars, which is usually completely full during working hours.

While access for cars to the central part of the campus is limited, two city streets intersect the rest of the campus, creating a serious burden on the transport infrastructure since they are often used for parking both by SPbPU visitors and by residents of nearby houses and employees of nearby enterprises. Parking regulations on these streets are determined by city authorities (the Committee for Transport Infrastructure of St. Petersburg), so any changes to this regulation must be agreed with them.

Communal Infrastructure

SPbPU ensures the maintenance and development of the communal infrastructure of its campus. The scale of this task can be demonstrated by the following figures:

- The length of external serviced heating networks is 20 km;
- The length of outdoor networks of drinking and fire water supply and sewerage systems is 30 km;
- The length of internal heat supply, water supply and sewerage systems more than 150 km. The number of heat nodes and water-measuring units more than 140;
- The number of sewers and water wells more than 800.

80% of all external and internal water and heat supply and sewerage systems have physical wear or exceeded service life that can cause utility accidents. This necessitates large-scale repair work, which will inevitably affect the transport infrastructure in the campus.

Electricity Supply Sub System

There are no electricity generators on the SPbPU campus. Power supply to SPbPU is carried out centrally from the city power grid. Small solar and wind generators are used only for testing and research. The following technical objects of the electricity supply are located at SPbPU campus:

- 24 electricity transformer substations,
- 42 transformers,
- caЬle lines 6kV with a length of around 25 km,
- caЬle lines 0,3kV with a length of around 50 km.

74% of the electricity transformers and 90% of cable lines at SPbPU campus are more than 45 years old while their lifetime is 20-25 years. All this significantly limits the possibilities for developing additional distributed electricity consumers on campus, such as charging stations for electric vehicles.

Although formally SPbPU is not the owner of the campus territory and real estate objects, it is obliged to ensure the development, maintenance and operation of the campus as a whole. It means that various technical solutions, from the construction of new buildings on campus to the development of mobility on its territory, mast be considered as elements of a technical subsystem and must develop in a coordinated manner.

Development of the Corporate Mobility Management System

Starting Point (the Problem Definition)

The strategic documents reflecting the vision and development directions of the SPbPU campus states that "Polytechnic University is a high-quality green habitat." This vision was formulated in frame of the BCR INTERREG project AREA21 that was mostly focused on holistic approach to energy efficiency of dormitories and teaching buildings in the campus. In another BCR INTERREG project LUCIA, this vision was extended to the smart energy efficient outdoor lighting system at the campus. At the same time, issues connected to development of corporate transport in relation to SPbPU campus remained outside the scope of implemented projects. These issues are among the key ones for achieving the formulated strategic goals for the development of the SPbPU campus, since transport systems have a significant impact on the implementation of sustainable development goals and the green agenda.

The main parameters of the SPbPU campus from the point of view of a corporate transport system are the following.

The fleet of vehicles owned by SPbPU consists of 6 buses, 24 cars and 11 units of special vehicles. Of the 120 hectares of campus, about 1 hectare is occupied by equipped parking lots.

On average, about 500 cars enter the campus every day, including cars of SPbPU employees, cars of residents whose apartments are located in 3 residential houses on campus, and cars belonging to SPbPU partners and service companies.

Currently, there is no system for detailed analysis of traffic on campus and parking management, and accounting for incoming vehicles is carried out only from the standpoint of checking permission to enter.

At present, a large number of personal vehicles of the SPbPU students and staff are parked in the adjacent streets or "wild" parking lots nearby, which creates a serious burden on the road services and the environment

Corporate Mobility System Development Targets

SPbPU has plans for the further development of the campus, which include both changes to the existing sites and the creation of a new Technopolis site on an area of 15 hectares. In these plans, serious attention is paid to transport management issues (open and underground parking, smart system for monitoring and management of corporate transport).

The main goals of this project are:

- minimizing the negative impact of transport on the historical buildings and park located on the territory of the central campus;
- reducing the number of cars parked in unregulated yard parking lots and unauthorized parking;
- encouraging the use of electric vehicles by SPbPU employees and students;
- encouraging the use of bicycles and similar environmentally friendly vehicles for travel between buildings and to/from the campus;
- attracting the attention and involvement of SPbPU employees and students in the complex problem of corporate transport development in SPbPU and increasing their motivation for green mobility;
- improving the efficiency of the corporate transport management system in SPbPU.

To achieve the set goals, the following activities were implemented during the pilot project:

- development of a concept for creating an intelligent corporate transport system on the SPbPU campus based on the best EU practices and with the broad participation of students, staff and other stakeholders;
- pilot implementation of individual provisions of the concept at designated sites on the campus (new parking on the existing site for 1000 spaces with protection from unauthorized parking nearby, new guarded parking for bicycles and similar eco-friendly vehicles, intelligent control and monitoring system for the corporate transport system, targeted measures to stimulate the use of eco-friendly mobility by SPbPU students and staff);
- analysis of feedback from the pilot sites for further development of the concept of the SPbPU corporate transport system and its promotion for use in urban planning of similar facilities.

There is no future without the past. Innovative development must be combined with respect for cultural and historical heritage, even if this requires additional efforts and resources or restrictions on the use of heritage sites. SPbPU carefully treats its history and does a lot to preserve the cultural and historical heritage. Unique historical buildings, a beautiful park in the center of the campus - all this is not only protected by the state, but is also the subject of close attention both in the process of operation and in the development of new projects. This fully applies to the project of development of the corporate mobility system at SPbPU . All technical solutions for lighting roads on campus, placement of control and security system elements, new parking lots, etc. are maintained in the historical style. If the cost or technical parameters of these solutions conflict with the requirements of compliance with the historical style, preference is given to the latter.

Target Groups of the Corporate Mobility System (Socio-Technical System: People First)

The project of creating a smart corporate transport system at SPbPU campus is based on the best world experience with the broad participation of students, employees and other stakeholders.

The target groups of the SPbPU CMMS development project are grouped into the following general categories:

(i) Local and regional authorities responsible for urban mobility as well as for the development of the transport system and the city as a whole. This includes policy makers as well as decision-making officials from various Committees of the Government of St. Petersburg (the Committee for the Development of Transport Infrastructure, the Committee for Urban Development and Architecture, the Committee for Nature Management, Environmental Protection and Environmental Safety of the Government of St. Petersburg). In addition, this target group also includes representatives of federal authorities, and in particular, the Ministry of Science and Education, which formally owns the territory and real estate in SPbPU.

(ii) Urban transport experts and researchers - technical experts, designers, mobility managers. This target group includes both experts from SPbPU and the regional authorities as well as from equipment and services suppliers and research organizations.

(iii) Users of urban mobility. The central campus of SPbPU is a point of attraction for cultural life for local residents. Its historical buildings often host classical and modern music concerts and other cultural events. SPbPU Park is a popular place for walks and recreation not only for SPbPU students, but also for city residents. Therefore, in addition to SPbPU students and staff, one of the main target audiences of the SPbPU SMMS are local residents.

(iv) SPbPU teachers and students. The project to create a corporate mobility system generates many real topics for student research and diploma theses. Using its results in the educational process improves the quality of education and the degree of involvement of students and teachers in strategic projects of SPbPU.

Efficiency of the Corporate Mobility Management System (Economic Subsystem)

The traditional approach to assessing the efficiency of any project is based on comparing the amount of investment in the project implementation and the necessary costs for using its results on the one hand, and direct and indirect income from the project on the other. For the project to create a corporate mobility system at SPbPU, this approach has limited application. This is due to the very nature of the system being created: its main goal is not to make a profit, but to improve the quality of the living environment on the SPbPU campus. Therefore, the task of ensuring the economic efficiency of the project to create corporate transport mobility at SPbPU is limited to issues of partial compensation of operating costs by users. Direct income from using the corporate transport mobility system can consist of tariffs for the entry of vehicles into the campus, as well as from the payment for additional services for users of the system (parking for electric vehicles with the possibility of recharging, rent of corporate means of individual mobility, etc.). However, paid entry to the company's territory for employees may cause natural bewilderment: why should I pay the employer for the opportunity to come to the workplace? The sociotechnical approach (people first) assumes taking into account such psychological aspects, even to the detriment of direct economic benefit.

For companies like SPbPU, a solution to the problem may be a multi-level system of tariffs for using the system:

- Free entry to the territory for people with disabilities;
- Preferential tariff for company employees and representatives of partner organizations (tenants and owners of premises on the company's territory, contractors, suppliers of goods and services, etc.);
- Full tariff for third-party users.

The payment procedures for using the corporate mobility system by the company's employees deserve special consideration. Let's imagine an employee of the company whose only income is the salary he/she receives from the company. At the same time, this salary is taxed in favor of the state. In Russia, such a tax is about 43%. If an employee transfers 500 rubles a month to the company for using the corporate mobility system, about 200 rubles go in taxes. For SPbPU, the number of users of the system is estimated at 2,000 people. This means that when they pay for access to the corporate transport system, taxes amount to about 400,000 rubles a month.

Now let's imagine that, with the employee's consent, the company reduces his/her salary by the same 500 rubles, but provides him/her with free access to the system. On a company scale, this can result in tax savings of up to 400,000 rubles a month.

Of course, this creates additional complications associated with the redistribution of company budget items, but an additional advantage of such a solution is an increase in the level of employee involvement in company projects, especially if the resulting savings are divided between the company and employees (for example, the preferential rate for them is halved).

Outdoor Lighting for the Corporate Mobility System Sustainability (Technical Subsystem)

Lighting is an important part of urban life and infrastructure, but at the same time, it is a huge consumer of energy. A significant portion of the energy consumed by the urban infrastructure is used for outdoor lighting. At the same time, outdoor lighting is the issue of an urban life quality. If a road, street or path in a park is not well lit at night, these areas are poorly suited for walking, cycling or e-scootering. This is due to both the limitations of human ability to see and navigate in the dark and the feeling of insecurity that many people often have in a dark area away from light sources. For SPbPU, this is particularly relevant, as the park in the center of the campus is not only a place for walking, but also a local route for moving between buildings. The issues of increasing comfort and safety by lighting roads and footpaths are closely related to the issues of economic efficiency. This efficiency depends on two main factors: the specific consumption of light sources and the operating mode of the system. LED solutions have created a great potential for significant energy savings for lighting. The simplest outdoor control systems have two operating modes: on during daylight hours (by time of day or by light sensors) and off during darkness. Intelligent outdoor lighting control systems also include motion sensors and dimming. When there are no people on the road, they ensure that the lighting is completely switched off or reduced in level to save energy. Motion sensors could make it more energy efficient and avoid lighting pollution (especially in green areas such as SPbPU park with high biodiversity). Another solution is do dim the light during the night hours to avoid light pollution. When lighting up a green area, it is important to have in mind the surrounding natural environment and make the lighting 'fit' the area. This could be done by downward facing lights, so only the path is highlighted, and the surrounding area is in its 'natural' dark phase. All this once again underlines the need to apply a holistic approach in planning and implementing projects for the development of sustainable CMMS.

RESULTS

Pilot implementation of the corporate mobility system on dedicated sites at the SPbPU campus has provided new parking in existing site on 1000 spots with protection from illegal parking nearby, new protected parking for bikes and similar green vehicles, smart system for management and monitoring of the corporate transport system, target measures for encourage the use of green mobility by SPbPSU students and staff.

The CMMS development project produce following outcomes:

1. Knowledge of best practices for locating parking lots next to landmark monuments, parks and buildings. This knowledge will expand the knowledge base of experts in St. Petersburg, which is rightfully considered the cultural capital of Russia and the central part of which is a single object of historical and cultural heritage. This knowledge will be used by SPbPU experts in projects for further development of campus area and will be transferred to the city urban planners and other stakeholders.
2. Knowledge of economic aspects of the transport component in the corporate management system of universities and similar organizations with large distributed territories and a large number of employees using personal transport.
3. Knowledge of holistic nature of a corporative transport system. This knowledge will be used for modernization of existing educational programs for the SPbPU students.
4. Feedback from the pilot project for further development of the SPbPU corporate transport system and for its promotion for use in urban planning for similar objects.

LESSONS LEARNED

1. The effectiveness of infrastructure projects, including urban mobility development projects, is very difficult, and sometimes impossible, to adequately assess solely using economic indicators (costs, income, profit, etc.). These indicators can be used at the project planning stage when assessing its scale and finding the necessary financial resources, when interacting with stakeholders, etc. This is especially relevant for infrastructure projects in the public sphere, since in this case end users do not pay at all or pay only partially for the benefits they consume using the created infrastructure (efficient outdoor lighting, convenient and safe roads, availability of services, etc.). This contradiction can only be resolved by including infrastructure projects in urban mobility development

strategies, and these strategies must be coordinated at the corporate, regional, and national levels.

2. Any infrastructure project requires careful preliminary coordination with other infrastructure or construction works. Inconsistency of works on different projects can lead to unnecessary costs, when, for example, a recently repaired road surface has to be destroyed in order to repair underground communications. This additionally confirms the need to implement projects for the creation of corporate transport systems within the framework of a single development strategy of the organization.
3. Successfully implemented infrastructure projects, such as corporate mobility development projects, can be an effective tool for increasing the global competitiveness of an organization. At SPbPU, this effect was demonstrated by increasing the level of student satisfaction with the quality of life on campus, which serves as a good argument in the competitive struggle for future applicants.
4. Attracting and involving users in an infrastructure project is a well-known way to increase its efficiency. For universities, including SPbPU, this involvement can be organized in the form of student research at various levels. Considering that SPbPU has the Institute of Machinery, Materials, and Transport, many specific issues of implementing the project to create a corporate transport system were used in organizing student research and graduation projects. Using the materials of the implemented project in the educational process, the results of which students can evaluate daily at their own level, improves the quality of training in both engineering disciplines related to transport and management disciplines.
5. The implementation of any infrastructure project at cultural heritage sites creates additional challenges associated with the presence of additional control by government authorities. This increases the risks, the volume of work required and its cost. These circumstances must be taken into account when planning such projects, in which careful attitude to historical or cultural objects takes priority over the desire to use the best and most effective innovative approaches and solutions.

CONCLUSION

For the successful implementation of projects for the creation and development of a corporate transport system, a holistic approach to the analysis and synthesis of the system based on a socio-technical approach is necessary.

The socio-technical approach to building a sustainable corporate mobility management system provides an adequate response to the current problems and challenges in the development of urban transport systems. These problems and challenges are typical for large cities. They became especially evident during the pandemic, but many of the organizational and technical solutions that emerged during this time remain relevant and are being further developed today.

The socio-technical approach to managing corporate and urban mobility is especially effective in companies, as well as organizations and areas of activity that generate significant passenger traffic and are usually difficult to plan for public transport:

- large companies and other large employers with a large number of passengers who get to a certain place at the same time of day;
- non-profit associations of policy makers, such as "Chambers of Commerce", as public voices, multipliers and future advisors for their members;
- industrial and commercial areas, usually located on the outskirts of cities and deprived of public transport infrastructure, in which several enterprises and a large number of jobs are concentrated;
- "sleeping areas" of large cities, where a large number of residents are concentrated in a relatively small area, who go to work in other areas of the city every morning, and return back after the end of the working day, thereby creating a large load on the city transport system;
- professions with irregular working hours or requiring the personal presence of workers at the workplace outside of "classic" working hours (for example, shift work, healthcare, retail, hotel business).

This is where a sociotechnical approach to managing corporate mobility can have the greatest impact, both at the company or organization level, and at the metropolitan level.

If local authorities, industrial and commercial companies as well as research organizations work together on local mobility planning, all actors involved will benefit from it. Cities can gain a greener mobility and better quality of life, companies a better image and lowered carbon footprint, and citizens a less stressful and cheaper commute to work. To ensure their sustainable uptake, the solutions need to be specific to the company, geographical area and professional sector.

REFERENCES

Adewale, T., & Paul, J. (2024). [and IoT: How These Technologies Are Creating the Perfect Storm for Smart Systems.]. *AI*, •••, 5G.

Albino, V., Berardi, U., & Dangelico, R. M. (2015). Smart cities: Definitions, dimensions, performance, and initiatives. *Journal of Urban Technology*, *22*(1), 3–21. DOI: 10.1080/10630732.2014.942092

Amara, E., Moussa-Jerbi, A., & Tanfous, F. (2024). *Stakeholder Dynamics in the Context of a Local Urban Project*. Social Business., DOI: 10.1362/204440824X170747761359874

Anke, J., Francke, A., Schaefer, L., & Petzoldt, T. (2021). Impact of SARS-CoV-2 on the mobility behaviour in Germany. *European Transport Research Review*, *13*(1), 10. Advance online publication. DOI: 10.1186/s12544-021-00469-3 PMID: 38624595

Brzeziński, Ł. (2024). Social Aspects of Smart Urban Mobility. *Encyclopedia.*, *4*(2), 864–873. DOI: 10.3390/encyclopedia4020055

Bylinko, L. (2024). Integration of an urban transportation system as an element of mobility plans. *Transport Problems*, *19*(1), 5–17. DOI: 10.20858/tp.2023.19.1.01

Challenger, R., & Clegg, C. W. (2013). Crowd disasters: a socio-technical systems perspective. ln: Drury, J., Stott, C.J.T. (Eds.), Crowds in the 21st century. Perspectives from contemporary social science. Routledge, Ablngdon, Oxfordshire OX14 4RN, United Kingdom, pp. 80-97

Coletta, C., Heaphy, L., & Kitchin, R. (2019). From the accidental to articulated smart city: The creation and work of 'Smart Dublin'. *European Urban and Regional Studies*, *26*(4), 349–364. DOI: 10.1177/0969776418785214

Correa, V. C., Lugo-Agudelo, L. H., Aguirre-Acevedo, D. C., Contreras, J. A. P., Borrero, A. M. P., Patiño-Lugo, D. F., & Valencia, D. A. C. (2020). Individual, health system, and contextual barriers and facilitators for the implementation of clinical practice guidelines: A systematic metareview. *Health Research Policy and Systems*, *18*(1), 74. DOI: 10.1186/s12961-020-00588-8 PMID: 32600417

Davis, M. C., Challenger, R., Jayewardene, D. N. W., & Clegg, C. W. (2014). Advancing sociotechnical systems thinking: A call for bravery. *Applied Ergonomics*, *45*(2), 171–18. DOI: 10.1016/j.apergo.2013.02.009 PMID: 23664481

Dediu, H. 2019. The micromobility definition. URL: https: //micromobility.io/news/the-micromobility-definition. Accessed: 2025-03-21

Fatima, H., & Rasool, M. (2025). AI-Driven Urban Mobility Solutions: Enhancing Smart Traffic Management and Urban Infrastructure Optimization. DOI: 10.13140/RG.2.2.25367.48803

Gheorghe, C., & Soica, A. (2025). Revolutionizing Urban Mobility: A Systematic Review of AI, IoT, and Predictive Analytics in Adaptive Traffic Control Systems for Road Networks. *Electronics (Basel)*, *14*(4), 719. DOI: 10.3390/electronics14040719

Graf, A., & Kantel, A. (2025). Planning sustainable and inclusive transformation: An energy democracy perspective on urban mobility. *Frontiers in Political Science*, *7*, 1421569. Advance online publication. DOI: 10.3389/fpos.2025.1421569

Gulc, A., & Budna, K. (2024). Classification of Smart and Sustainable Urban Mobility. *Energies*, *17*(9), 2148. DOI: 10.3390/en17092148

Heumann, M., Kraschewski, T., Otto, P., Tilch, L., Brauner, T., & Breitner, M. (2025). Factors influencing the usage of shared micromobility: Implications from Berlin. *Journal of Cycling and Micromobility Research.*, *100063*, 100063. Advance online publication. DOI: 10.1016/j.jcmr.2025.100063

Irvin, R. A., & Stansbury, J. (2004). Citizen participation in decision making: Is it worth the effort? *Public Administration Review*, *64*(1), 55–65. DOI: 10.1111/j.1540-6210.2004.00346.x

Leavitt, H. J. (1965). Applied Organizational Change in Industry: Structural, Technological, and Humanistic Approaches. In March, J. G. (Ed.), *Handbook of Organizations, 1st* (pp. 1144–1170). Routledge.

Ma, C. & Vu, H. & Wang, J. & Trieu, V.H. & Li, G. (2023). Understanding Residents' Behavior for Smart City Management by Sequential and Periodic Pattern Mining. IEEE Transactions on Computational Social Systems. PP. 1-17. DOI: 10.1109/TCSS.2023.3249740

Mkrtchyan, N. (2024). Drawing of the Russian population into large cities and their suburbs. *Journal of the New Economic Association, New Economic Association*, *63*(2), 241–248. DOI: 10.31737/22212264_2024_2_241-248

Moser, D. (2021). Approaches to transformative urban mobility. *Journal of Urban Regeneration and Renewal*, *14*(3), 255. Advance online publication. DOI: 10.69554/CNPJ9514

Mykhailova, K. (2023). *Implementation of urban mobility projects for smart cities: approaches to justification and evaluation*. Black Sea Economic Studies., DOI: 10.32782/bses.84-20

Narayanan, S., & Antoniou, C. (2023). Shared mobility services towards Mobility as a Service (MaaS): What, who and when? *Transportation Research Part A, Policy and Practice*, *168*, 103581. Advance online publication. DOI: 10.1016/j.tra.2023.103581

Nurulin, Y., Skvortsova, I., & Kaltchenko, O. Charging Infrastructure for Electric Vehicles: Problems and Development Prospects. Lecture Notes in Networks and Systems, 2021, 460 LNNSP, pp. 239–248. DOI: 10.1007/978-3-031-20875-1_22

Pawlowska, Barbara & Suchanek, Michał. (2019). Transport as a Factor in the Achievement of the EU Goals to Combat Climate Changes and to Reduce Greenhouse Gases Emissions. Journal of Management and Financial Sciences. 79-96. DOI: 10.33119/JMFS.2018.33.5

Rajkumar, S., Sandhiya, S., & Manjusri, A. (2025). Optimized Traffic Routing System for Urban Congestion Management. *Journal of Ubiquitous Computing and Communication Technologies.*, *7*(1), 19–35. DOI: 10.36548/jucct.2025.1.002

Rodrigues, J.-P. (2024). *The Geography of Transport Systems* (6th ed.). Routledge., DOI: 10.4324/9781003343196

Sankhwar, P. (2024). Suitability of Electric Vehicle Charging Infrastructure and Impact on Power Grid Due to Electrification of Roadway Transportation with Electric Vehicles. [IJSR]. *International Journal of Science and Research (Raipur, India)*, *13*(12), 374–383. DOI: 10.21275/SR241204033720

Sluiter, C. (2021). The new EU Sustainable and Smart Mobility Strategy: A local and regional perspective https://www.cidob.org/sites/default/files/2024-07/25-30_CASPAR%20SLUITER_ANG.pdf

Sony, M., & Naik, S. (2020). Industry 4.0 integration with socio-technical systems theory: A systematic review and proposed theoretical mode. *Technology in Society*, *61*, 10124. DOI: 10.1016/j.techsoc.2020.101248

Suvvari, S., & Saxena, V. D. (2023). Stakeholder Management in Projects: Strategies for Effective Communication. *Innovative Research Thoughts.*, *9*(5), 188–201. DOI: 10.36676/irt.v9.i5.1479

Szpilko, D., Budna, K., & Drmeyan, H. Remiszewska, (2023). Sustainable and smart mobility—Research directions: A systematic literature review. Ekon. Sr, 3, 31–61. ttps://DOI: <ALIGNMENT.qj></ALIGNMENT>10.34659/eis.2023.86.3.584

The World Bank. Urban Development. 2017. Available online: https://www.worldbank.org/en/topic/urbandevelopment/overview#:~:text=Today%2C%20more%20than%20half%20of,people%20will%20live%20in%20cities. (accessed on 5 January 2024)

Tuli, F. M., Mitra, S., & Crews, M. B. (2021). Factors influencing the usage of shared e-scooters in Chicago. *Transportation Research Part A, Policy and Practice, 154*, 164–185. DOI: 10.1016/j.tra.2021.10.008

Uribe-Chavert, P., Posadas-Yagüe, J.-L., & Poza-Luján, J.-L. (2025). Evaluating Traffic Control Parameters: From Efficiency to Sustainable Development. *Smart Cities.*, *8*(2), 57. DOI: 10.3390/smartcities8020057

Vorwerg, F., Torkayesh, A., & Venghaus, S. (2025). From fossil fuels to alternative fuels: Strategy development for a sustainable transport sector in Germany. *Energy, Sustainability and Society*, *15*(1), 4. Advance online publication. DOI: 10.1186/s13705-024-00498-5

Wang, L. & Shang, A. & Feng, Y. & Fu, H. & Gao, Z. (2025). Research on Regional Traffic Simulation Analysis and Traffic Capacity Improvement Technology Under the Background of Urban Renewal. E3S Web of Conferences. 617. DOI: 10.1051/e3sconf/202561702001

Wolniak, R. (2023). Smart mobility in a smart city concept. Scientific Papers of Silesian University of Technology. *Organization and Management Series.*, *2023*(170), 679–692. Advance online publication. DOI: 10.29119/1641-3466.2023.170.41

Chapter 6
Emerging Technologies and Shift in Employment in the Transportation Sector

Satya Subrahmanyam
https://orcid.org/0000-0003-0441-2742
Holy Spirit University of Kaslik, Lebanon

ABSTRACT

The transportation sector is undergoing a profound transformation driven by emerging technologies such as autonomous vehicles, electrification, IoT, artificial intelligence, blockchain, and robotics. This chapter examines how these advancements are re-shaping employment, introducing new technology-focused roles while displacing traditional ones. It delves into automation's impact on workforce dynamics, the rise of gig-based employment, and the skillsets required for future opportunities. Ethical and social challenges, including job security and economic disparities, are analyzed alongside strategies for fostering resilience through reskilling and adaptive workforce models. Case studies and policy recommendations highlight the collaborative roles of governments, corporations, and educational institutions in achieving equitable workforce transitions, ensuring that technological progress aligns with social responsibility and sustainable employment strategies.

DOI: 10.4018/979-8-3373-0882-1.ch006

1. INTRODUCTION

The evolution of urban mobility stands at a critical intersection of technological innovation and sustainability imperatives. As cities around the world grapple with the challenges of rapid urbanization, population growth, and climate change, the transportation sector emerges as both a contributor to and a potential solution for achieving sustainable development goals. Traditional modes of transportation, heavily reliant on fossil fuels and manual operations, are increasingly being replaced or augmented by digital and automated technologies that promise greater efficiency, safety, and environmental stewardship. This transformation is not only reshaping how people and goods move within cities but also redefining the structure of employment in the transportation sector.

1.1 Contextualizing Urban Mobility and Transportation Trends

Urban mobility has evolved from being merely a logistical concern to becoming a central component of sustainable urban planning. Modern cities are prioritizing integrated, multimodal transport systems that align with broader environmental and social objectives. The global push for reducing greenhouse gas emissions has placed transportation—responsible for nearly 24% of direct CO_2 emissions from fuel combustion—under increased scrutiny (International Energy Agency [IEA], 2022). Consequently, governments, industry stakeholders, and urban planners are investing in public transit, cycling infrastructure, and digital platforms that promote shared and eco-friendly mobility options.

At the same time, the proliferation of on-demand services, the emergence of mobility-as-a-service (MaaS) platforms, and the growing preference for micro-mobility solutions like e-scooters and bike-sharing schemes indicate a shift in consumer behavior. This transformation is driven not only by environmental consciousness but also by the demand for convenience and real-time responsiveness. As a result, cities are becoming testbeds for innovative mobility solutions that integrate data analytics, artificial intelligence (AI), and the Internet of Things (IoT) into everyday transportation systems (Subrahmanyam, 2025).

1.2 Importance of Technological Advancement in Sustainable Transport

Technology is playing an indispensable role in enabling sustainable transport by reducing dependency on fossil fuels, enhancing system efficiency, and improving user experience. The development of electric vehicles (EVs), autonomous driving systems, and smart traffic management solutions illustrate how innovation can align with environmental objectives. The deployment of AI-based predictive traffic modeling helps mitigate congestion, thereby lowering emissions and fuel consumption (Subrahmanyam et al., 2024). Autonomous vehicles (AVs) promise not only to transform mobility but also to reduce accidents caused by human error and to optimize road usage.

In addition to these environmental and efficient gains, emerging technologies also support inclusive transportation. Smart apps designed with accessibility in mind can improve travel for individuals with disabilities, while data-driven route planning can address service gaps in underserved areas. Importantly, technological innovation also facilitates the decarbonization of logistics networks through the integration of electric trucks, drone deliveries, and warehouse automation—all of which contribute to lowering the overall carbon footprint of urban freight movement (Hilmola & Henttu, 2022).

Yet, as these technological advancements create new opportunities, they also disrupt established systems and job roles. Drivers, dispatchers, maintenance workers, and administrative personnel face increasing uncertainty about their future roles as automation becomes more prevalent. Thus, while emerging technologies are vital to sustainable urban transport, they simultaneously necessitate careful consideration of their socio-economic impacts, particularly regarding employment.

1.3 Objective and Scope of the Chapter

This chapter seeks to explore the dynamic interplay between emerging technologies and the evolving employment landscape within the transportation sector, particularly in the context of sustainable urban mobility. It investigates how innovations such as automation, electrification, AI, IoT, and smart infrastructure are not only transforming the technical architecture of transport systems but also reshaping job roles, skill demands, and workforce demographics.

The chapter is structured to provide a comprehensive analysis of both the positive and adverse effects of technological integration. It begins with an overview of key emerging technologies influencing the transportation domain, followed by an examination of their impact on employment—highlighting job displacement, creation, and transformation. It incorporates case studies from different regions to illustrate global trends and localized responses to these shifts. The chapter further evaluates how these changes intersect with sustainability goals, including the promotion of green jobs and equitable access to employment.

The chapter discusses the policy and educational interventions necessary to facilitate a just transition for workers affected by technological change. It emphasizes the need for collaborative efforts among governments, industry, and academic institutions to develop strategies that balance innovation with social responsibility.

The chapter positions the shift in transportation employment not merely as a consequence of technological disruption but as a critical dimension of sustainable urban development. By understanding and proactively addressing these changes, stakeholders can ensure that the future of transportation is not only smarter and greener but also more inclusive and equitable.

2. OVERVIEW OF EMERGING TECHNOLOGIES IN THE TRANSPORTATION SECTOR

The transportation sector is undergoing a profound transformation driven by a convergence of emerging technologies. These innovations not only enhance operational efficiency and reduce environmental impact but also reshape mobility patterns, infrastructure requirements, and labor demands. This section examines five critical technological advancements—autonomous vehicles (AVs), electrification and battery technologies, the Internet of Things (IoT), artificial intelligence (AI) and machine learning (ML), and robotics and drones—and explores their implications for the transportation sector.

2.1 Autonomous Vehicles (AVs)

One of the most revolutionary transportation innovations of the current day is the autonomous car. For navigation and operation, they use a variety of sensors, camera, radar, lidar, GPS, and onboard computer systems. The amount of human interaction required varies. Level 5 fully automates all aspects of an AV, leaving no room for human intervention, while Level 0 completely lacks automation, according to the Society of Automotive Engineers (SAE) (SAE International, 2021). Level 2 (partially automated with driver monitoring) is where the majority of commercially available cars are at the moment, although Levels 4 and 5 are rapidly approaching.

Implications for Public and Private Transport Services - The deployment of AVs is expected to revolutionize both public and private transport services. In the public sector, autonomous buses and shuttles can enhance the frequency and reliability of transit services, especially in underserved or low-density areas. Pilot projects in cities such as Helsinki, Singapore, and Phoenix have demonstrated the potential of AVs to fill first-mile and last-mile connectivity gaps (Milakis et al., 2017). In the private sector, AVs are transforming ride-hailing services. Corporates like Waymo and Tesla are investing heavily in robo-taxi models, which could significantly reduce transportation costs and increase accessibility.

The widespread adoption of AVs also raises concerns about employment displacement, especially for drivers in taxis, freight, and public transit services. Regulatory frameworks, infrastructure upgrades, and ethical considerations (e.g., decision-making in emergency situations) must also be addressed before full-scale integration is realized (Cugurullo, 2020).

2.2 Electrification and Battery Technologies

An essential component of worldwide plans to reduce carbon emissions is the electrification of automobiles. Emissions of greenhouse gases and air pollution are drastically reduced by electric cars (EVs) driven by modern lithium-ion or solid-state batteries since these vehicles do not emit any emissions from their tailpipes. Worldwide, more than 26 million electric vehicles were in use in 2023, and the IEA predicts robust growth in the next decade (IEA, 2023).

Rise of EVs and Associated Infrastructure - Governments worldwide are incentivizing EV adoption through subsidies, tax breaks, and investments in charging infrastructure. High-capacity fast-charging stations, wireless charging, and battery swapping stations are being introduced in urban areas to alleviate range anxiety and support EV scalability. Smart grids are also being developed to manage energy loads efficiently and integrate renewable sources (Subrahmanyam et al., 2024).

Impacts on Traditional Automotive Manufacturing - The shift to EVs necessitates a fundamental reconfiguration of automotive manufacturing. Unlike internal combustion engine (ICE) vehicles, EVs have fewer moving parts, altering the requirements for parts suppliers and maintenance services. Traditional automakers are investing in retraining their workforce, repurposing production lines, and establishing battery manufacturing facilities. This transition creates opportunities in green jobs but simultaneously threatens the livelihoods of workers in ICE-related supply chains (Maurya et al., 2025).

2.3 Internet of Things (IoT) and Smart Infrastructure

The integration of IoT into transportation systems is facilitating the development of smart infrastructure and connected mobility ecosystems. IoT involves embedding sensors and communication devices into vehicles, roads, traffic lights, and other urban infrastructure to collect and exchange real-time data.

Connected Vehicles and Smart Traffic Systems - Connected vehicles (V2V – vehicle-to-vehicle, and V2I – vehicle-to-infrastructure) communicate with each other and with urban traffic systems to improve safety and traffic flow. For example, real-time traffic data can be used to reroute vehicles to avoid congestion, reduce idle times, and lower emissions. Smart traffic lights can adjust signal timing dynamically based on vehicle density, enhancing efficiency (Hussain et al., 2020).

IoT-enabled predictive maintenance helps fleet operators monitor vehicle health remotely and schedule repairs before breakdowns occur. This reduces operational costs and enhances safety. Cities like Barcelona and Amsterdam are leading in deploying smart mobility solutions, using IoT to manage parking, public transit, and road maintenance more effectively (Batty et al., 2012).

2.4 Artificial Intelligence and Machine Learning

The digital revolution of transportation is built upon artificial intelligence and machine learning. To aid in decision-making, boost operational efficiency, and enhance the user experience, these technologies handle massive amounts of data.

Predictive Analytics, Route Optimization, and Fleet Management - AI-driven predictive analytics enables transport systems to anticipate traffic conditions, vehicle performance, and passenger demand. ML algorithms can analyze historical traffic data and weather forecasts to optimize routing and reduce travel time (Zhao et al., 2019). In logistics, AI optimizes delivery routes in real-time, accounting for traffic conditions, delivery windows, and vehicle load capacities.

Fleet management systems benefit greatly from AI by monitoring vehicle locations, driver behavior, fuel consumption, and maintenance needs. Companies like UPS and DHL use AI to enhance delivery efficiency and minimize costs. In public transport, AI supports dynamic scheduling and real-time passenger information systems, improving service reliability and passenger satisfaction (Batarseh & Freeman, 2021).

AI plays a pivotal role in enabling autonomous driving technologies by helping vehicles interpret sensor data and make split-second decisions. Natural language processing (NLP) and computer vision also support voice-controlled navigation and object recognition in AVs.

2.5 Robotics and Drones

The use of robotics and drones is rapidly expanding in logistics and transportation, particularly in the areas of warehouse automation and last-mile delivery.

Automation in Logistics and Last-Mile Delivery - Robotic process automation (RPA) and physical robots are revolutionizing warehousing and distribution. Automated guided vehicles (AGVs), robotic arms, and sorting systems increase efficiency and reduce human error in inventory management and order fulfillment. Amazon, for example, has deployed thousands of robots in its fulfillment centers, allowing for faster processing of online orders (Wurman et al., 2008).

Drones are being deployed for last-mile delivery in urban and rural settings, especially for high-value or time-sensitive goods such as medical supplies. Companies like Zipline and Wing have demonstrated the viability of drone delivery in both developed and developing regions. Drones can circumvent traffic congestion and reach remote areas inaccessible by traditional vehicles, enhancing delivery speed and reliability (Otto et al., 2018).

Drone integration faces challenges including airspace regulation, safety, payload limitations, and privacy concerns. Automation in logistics may displace low-skilled jobs, necessitating strategies for workforce transition and reskilling.

Emerging technologies are fundamentally reshaping the transportation sector by making mobility smarter, cleaner, and more efficient. Autonomous vehicles, electrification, IoT, AI, and robotics are not isolated innovations but interconnected forces driving systemic change. These technologies enhance sustainability and urban mobility but also require thoughtful integration to ensure that social and economic disruptions—particularly those related to employment—are minimized. As the sector continues to evolve, a balanced approach involving technological advancement, policy development, and human capital investment will be critical to harness the full potential of innovation in transportation.

3. TRANSFORMATION OF EMPLOYMENT STRUCTURES IN THE TRANSPORTATION SECTOR

Automated vehicles (AVs), electric vehicles (EVs), IoT, artificial intelligence (AI), and robotics are some of the new technologies that are changing the transportation industry. Strong reskilling programs will be required as a result of these innovations' far-reaching effects on employment structures, which are changing skill needs, job profiles, and the nature of work itself. Job loss, job gain, shifting skill needs, and the rising significance of reskilling and upskilling are four important perspectives from which to examine this change.

3.1 Job Displacement

One of the most immediate concerns associated with technological disruption in transportation is the potential for job displacement. The automation of driving tasks through AVs threatens the livelihood of millions employed as professional drivers. A report by McKinsey & Company (2022) projected that by 2030, AVs could reduce the need for drivers in trucking, taxis, and ride-hailing services by up to 50%. Traditional vehicle mechanics may see a decline in demand, as EVs require less frequent service and possess fewer mechanical components compared to internal combustion engine vehicles (IEA, 2023).

Dispatchers and other logistics coordinators are also facing automation challenges due to AI-driven fleet management systems and smart logistics platforms that can autonomously handle route planning and delivery scheduling (Manyika et al., 2017). These shifts suggest a gradual erosion of routine and manual roles within the transportation sector, underscoring the urgency for proactive workforce planning.

3.2 Job Creation

While some traditional roles are being phased out, emerging technologies are simultaneously generating new employment opportunities. The transition to a technology-driven transportation ecosystem requires a new cadre of professionals equipped with technical expertise. The rise of AVs has fueled demand for AV software engineers, sensor calibration specialists, and remote vehicle operators (Smith & Anderson, 2021).

EVs has created roles in battery technology, charging infrastructure development, and EV maintenance. With the proliferation of connected and smart transport systems, cybersecurity experts and data analysts are in high demand to protect critical infrastructure and derive actionable insights from transportation data (Ghosh & Choi, 2022). These emerging roles are often higher-paying and require a deeper knowledge base, thus contributing to the evolution of job hierarchies within the sector.

3.3 Changing Skill Requirements

As the transportation sector becomes increasingly digitized and automated, the skill sets required to thrive in the industry are undergoing a substantial transformation. Traditional mechanical and manual skills, though still relevant in some contexts, are being supplemented—and in some cases supplanted—by digital competencies. Workers are now expected to possess at least basic digital literacy, as many routine tasks become integrated with digital tools and platforms.

STEM (science, technology, engineering, and mathematics) skills are becoming indispensable, particularly for roles involving system development, data analysis, and smart infrastructure management. A study by the World Economic Forum (2020) emphasized that adaptability, critical thinking, and complex problem-solving will be among the most sought-after skills in future transportation careers. Soft skills such as communication, collaboration, and emotional intelligence remain important, particularly in roles involving human oversight of autonomous systems or customer service in tech-enabled environments.

3.4 Role of Reskilling and Upskilling

Given the rapid pace of technological change, continuous learning and skill development have become essential for both individuals and organizations. Reskilling refers to training individuals for entirely new roles, while upskilling enhances existing skill sets to meet evolving job requirements. Both are critical to mitigating the negative employment impacts of digital disruption in transportation.

Industry-academia partnerships play a vital role in facilitating this transition. Universities and technical institutes are beginning to offer specialized programs in AV technology, electric mobility, and AI applications in logistics (OECD, 2022). At the same time, corporate-led initiatives such as internal training programs and collaborative platforms like Udacity's partnership with Mercedes-Benz for AV training demonstrate the private sector's commitment to workforce transformation (Udacity, 2021).

Governments also have a pivotal role to play in crafting inclusive labor policies that incentivize reskilling, support displaced workers, and promote equal access to educational resources. Public-private partnerships can help build a robust ecosystem that aligns education with industry needs, ensuring that workers are not left behind in the race toward digital transformation.

The transformation of employment structures in the transportation sector reflects the broader narrative of technological disruption and digital evolution. While automation and digital technologies pose challenges in the form of job displacement, they also offer vast potential for job creation and productivity enhancement. The key lies in proactively addressing these shifts by equipping the workforce with future-ready skills through focused reskilling and upskilling programs. Collaborative efforts among government, academia, and industry stakeholders are essential to ensure a smooth transition that balances innovation with inclusive workforce development.

4. CASE STUDIES AND GLOBAL TRENDS

The integration of emerging technologies into the transportation sector is unfolding differently across various regions, shaped by diverse policy priorities, labor market dynamics, and technological capabilities. This section explores key regional trends and case studies to illustrate how automation, electrification, and digital innovations are transforming employment structures in transportation. These global examples demonstrate both the opportunities and challenges inherent in aligning sustainable mobility goals with employment and workforce development.

4.1 North America: Trucking Automation and Employment

The advent of fully autonomous trucking systems is having a revolutionary effect on the transportation sector throughout North America, and the US in particular. Companies like Aurora, TuSimple, and Waymo Via are spearheading the rollout of Level 4 autonomous trucks, which can navigate roads partially or fully without human assistance (CB Insights, 2022). High operational expenses, increasing e-commerce demands, and ongoing driver shortages are propelling this trend towards automation.

While autonomous trucking promises enhanced efficiency, it raises concerns about the displacement of long-haul truck drivers, a sector that employs over 1.5 million people in the U.S. alone (Bureau of Labor Statistics, 2023). A study by the University of Pennsylvania projected that widespread adoption of AVs could reduce employment in long-haul trucking by up to 25% over the next two decades (Gittleman & Monaco, 2020). New roles are also emerging, including remote vehicle operators, AV safety engineers, and logistics software developers.

To address potential disruptions, North America has witnessed growing investments in retraining programs and public-private partnerships. For instance, the U.S. Department of Transportation has supported pilot programs aimed at preparing the workforce for digital logistics and transportation innovation (USDOT, 2022). These efforts signal an increasing recognition of the need to balance automation with inclusive workforce strategies.

4.2 Europe: Green Jobs and EV Transitions

Europe has positioned itself at the forefront of sustainable transportation, with a strong policy focusing on decarbonization and green employment. The European Green Deal and the Fit for 55 packages target a 100% reduction in CO_2 emissions from new cars by 2035, accelerating the transition to electric vehicles (European Commission, 2021). This shift is transforming employment in the automotive and transportation sectors, particularly in countries like Germany, France, and the Netherlands.

As combustion engine production phases out, traditional automotive jobs such as engine assembly and mechanical repair are declining. A study by the Fraunhofer Institute estimated that the German automotive sector could lose 100,000 jobs related to combustion technologies by 2030 (Weiss et al., 2020). This job loss is offset by the growth of green jobs, including EV battery production, charging infrastructure deployment, and renewable energy integration.

The EU's Just Transition Mechanism has facilitated skill development and labor mobility, ensuring that displaced workers are retrained for emerging roles in the green economy. Initiatives like the Automotive Skills Alliance and the European

Battery Alliance demonstrate coordinated efforts to align technological innovation with job creation and social sustainability (European Commission, 2021). These policies highlight Europe's proactive approach to managing the socio-economic impacts of the EV transition.

4.3 Asia: Smart Mobility Hubs and Ride-Hailing Ecosystems

In Asia, rapid urbanization and digital penetration have given rise to innovative smart mobility hubs and dynamic ride-hailing ecosystems. Cities like Singapore, Seoul, and Tokyo are investing in integrated transportation platforms that combine public transit, bike-sharing, and autonomous shuttles. These smart mobility systems use real-time data, AI, and IoT to optimize urban transportation and reduce congestion (World Bank, 2022).

Simultaneously, countries such as China and India have witnessed the exponential growth of ride-hailing services like Didi, Grab, and Ola, which have created millions of gig economy jobs. These jobs often lack formal protection, social security, or long-term stability. According to the International Labour Organization (ILO, 2021), while ride-hailing platforms have generated short-term employment, they have also contributed to income precarity and labor informality.

To address these concerns, countries like Singapore have implemented regulatory frameworks to support platform workers, offering social security benefits, insurance coverage, and training programs (Ministry of Manpower Singapore, 2021). Smart mobility initiatives are creating new roles in data management, mobility-as-a-service (MaaS) design, and urban traffic analytics, signaling a shift toward a more technology-oriented labor landscape.

4.4 Global Logistics: Drones and Automated Warehousing

The global logistics sector is at the cutting edge of automation, with innovations in drones and warehouse robotics redefining last-mile delivery and supply chain management. Corporates like Amazon, JD.com, and DHL are experimenting with drone deliveries, autonomous ground vehicles, and fully automated fulfillment centers (McKinsey & Company, 2023). These technologies promise to enhance speed, reduce human error, and cut operational costs.

Automated warehouses powered by AI and robotics can process orders more efficiently, but they also reduce demand for manual labor in packaging, sorting, and inventory management. In the U.S., Amazon's robotics expansion has displaced some warehouse roles while simultaneously creating new tech-oriented jobs such as robotics technicians and system engineers (Rosenblat & Stark, 2021). JD.com's smart warehouses in China have achieved high levels of automation, with a significantly leaner human workforce.

Despite the reduction in low-skill roles, the logistics sector is witnessing a demand spike for workers proficient in robotic systems maintenance, AI algorithms, and supply chain analytics. Vocational training centers and corporate academies are responding with upskilling programs tailored to logistics automation, reflecting a growing awareness of the sector's evolving skill requirements (OECD, 2022).

The global transformation of the transportation sector through emerging technologies reveals a complex interplay between innovation, employment, and policy. North America's focus on AVs, Europe's EV-led green jobs strategy, Asia's smart mobility platforms, and the global logistics revolution through automation each present unique labor market dynamics. While job displacement remains a legitimate concern, new opportunities are emerging in tech-centric and sustainable roles. Regional responses—from regulatory reform to upskilling initiatives—underscore the importance of inclusive, forward-looking employment strategies that align with technological advancement and sustainability goals.

5. IMPLICATIONS FOR URBAN MOBILITY AND SUSTAINABILITY

Emerging technologies in the transportation sector are not only reshaping employment structures and industrial operations but are also playing a transformative role in redefining urban mobility and sustainability. As cities strive to become smarter, greener, and more inclusive, technological innovations are integral in achieving sustainable development goals. This section explores the multifaceted implications of these technologies on public transport efficiency, sustainable employment models, and equity in access to mobility.

5.1 Efficient Public Transport Systems

The integration of artificial intelligence (AI), machine learning, and real-time data analytics into public transportation is revolutionizing how cities manage mobility. AI-based tools are increasingly used to optimize route planning, reduce traffic congestion, and lower greenhouse gas emissions. For instance, smart transit systems can analyze historical and real-time data to adjust bus or metro schedules based on passenger demand, weather conditions, and traffic flow (Mohamed et al., 2020). This dynamic routing capability helps minimize idle time, reduce fuel consumption, and improve punctuality.

Predictive analytics, a branch of AI, also enabling transit authorities to anticipate breakdowns or service disruptions through predictive maintenance, thus reducing downtime and enhancing system reliability (Sharma & Ghosh, 2021). Transport for London has employed AI to improve the scheduling of its bus fleet and reduce operational costs while enhancing service quality. Eco-routing algorithms have been developed to select the most fuel-efficient paths for vehicles, directly contributing to carbon emission reductions (Pojani & Stead, 2017).

With the rise of Mobility-as-a-Service (MaaS), urban centers are increasingly integrating various modes of transport—such as buses, trains, bikes, and ridesharing—into unified platforms that allow seamless, real-time trip planning and ticketing. These integrated platforms promote multimodal transport usage, reduce the reliance on private vehicles, and ultimately support low-emission urban environments.

5.2 Sustainable Employment Models

The shift toward sustainable transportation is also giving rise to green jobs that align economic growth with environmental stewardship. As cities transition to electric mobility and circular economy models, new employment opportunities are emerging in sectors such as renewable energy integration, EV maintenance, battery recycling, and sustainable logistics (International Labour Organization [ILO], 2018).

The growing demand for electric vehicles (EVs) has created a parallel demand for technicians skilled in battery systems, software integration, and electric drivetrains. Battery recycling plants and the refurbishment of vehicle components contribute to a circular economy, where resources are reused and waste is minimized (Chambers et al., 2020). These activities support the creation of stable, environmentally responsible jobs that reduce reliance on finite resources.

Sustainable employment models emphasize the importance of social inclusion. Public works programs focused on building cycling infrastructure or retrofitting diesel buses with cleaner technologies can be targeted toward vulnerable populations, such as low-income youth or displaced workers from legacy industries. These inclusive models of employment ensure that the benefits of green mobility are equitably distributed.

Beyond technical roles, sustainable transport ecosystems also require expertise in environmental impact assessment, urban planning, and policy design. As such, new interdisciplinary careers are forming that blend engineering, data science, and urban studies to support holistic transport development. Educational institutions and government bodies must, therefore, invest in reskilling programs and vocational training to prepare the workforce for this green transition.

5.3 Equity and Accessibility

While technological advancements offer tremendous potential for enhancing urban mobility, they also pose the risk of exacerbating existing socio-economic inequalities if not implemented thoughtfully. Digital transportation services, such as app-based ride-hailing, smart ticketing, and AI-based routing systems, often assume access to smartphones, internet connectivity, and digital literacy—resources that may not be universally available, particularly among the elderly, disabled, or low-income groups (Lucas, 2019).

The automation of public and private transport services could lead to job displacement in lower-skilled occupations, particularly affecting marginalized populations who rely on these roles for livelihood. Without inclusive labor policies and accessible training programs, the benefits of automation could be unevenly distributed, further widening socio-economic gaps.

To mitigate these risks, cities must adopt inclusive mobility frameworks that prioritize universal access. Transport innovations should be designed with features such as multilingual interfaces, offline functionality, and accessible formats for visually impaired users. Likewise, fare structures must remain affordable, and subsidies may be necessary to ensure that disadvantaged groups are not priced out of smart transport systems.

Urban planners should also ensure that smart mobility hubs are equitably distributed across neighborhoods, including underserved areas. This spatial equity is essential for enhancing social mobility and ensuring that technological shifts do not reinforce patterns of exclusion or segregation (Martens, 2017). Public engagement and participatory design processes can empower communities to co-create mobility solutions that reflect their needs and preferences.

Inclusive sustainability also involves addressing the needs of gender-diverse users. Women, for example, tend to rely more on public transportation and may have specific safety and caregiving-related needs that must be considered in the design and operation of transit systems (Shaw et al., 2020). Gender-sensitive planning and community outreach are critical for ensuring that emerging transportation technologies serve the entire urban population.

The evolving landscape of transportation technologies presents significant implications for urban mobility and sustainability. Through AI-powered public transport optimization, sustainable employment in green sectors, and inclusive access to digital mobility services, cities can align technological advancement with environmental and social goals. However, realizing these benefits requires careful planning, robust public policy, and cross-sectoral collaboration to ensure that no group is left behind in the transition to smart and sustainable urban mobility.

6. POLICY AND REGULATORY CONSIDERATIONS

As emerging technologies such as automation, electrification, and artificial intelligence transform the transportation sector, they also present complex policy and regulatory challenges. Governments, labor organizations, and international agencies must collaborate to develop adaptive frameworks that ensure fair labor practices, encourage sustainable technological adoption, and establish ethical standards that can operate across borders. This section explores the implications of these changes from three key policy perspectives: labor protections in the age of automation, incentives for green technology adoption, and the development of international standards and ethical governance.

6.1 Labor Regulations and Protection in the Age of Automation

Automation in transportation—through autonomous vehicles, robotics in logistics, and AI-based systems—has raised concerns regarding the displacement of traditional jobs. Long-haul truck drivers, delivery personnel, dispatchers, and maintenance workers are increasingly vulnerable to job loss as companies adopt cost-saving and efficiency-enhancing technologies (Frey & Osborne, 2017). In response, labor regulations must evolve to address the emerging risks associated with automation, while safeguarding workers' rights and enabling equitable transitions into new forms of employment.

Firstly, the redefinition of job classifications is essential. Many gig workers in ride-hailing platforms such as Uber or Lyft are categorized as independent contractors, lacking access to health insurance, retirement benefits, or job security (De Stefano, 2016). This classification often excludes them from traditional labor protections such as minimum wage laws and collective bargaining. Governments must establish clearer definitions and standards for employment in the digital economy to ensure fair compensation and protection for such workers.

Secondly, policies should incorporate just transition principles. These principles advocate for a holistic approach that combines social protection, retraining programs, and inclusive dialogue between employers, unions, and governments to support workers affected by structural changes in the labor market (ILO, 2018). Proactive reskilling and social safety nets can help displaced workers re-enter the workforce in emerging roles related to AI, cybersecurity, and green technologies in transport.

Labor laws should address workplace surveillance and data usage—issues becoming more prominent as AI and IoT-based tools are deployed to monitor worker productivity and behavior. Regulatory mechanisms are required to balance technological efficiency with individual privacy and autonomy, ensuring that algorithmic decision-making in workforce management does not reinforce biases or exploitative practices (Subrahmanyam, 2025).

6.2 Government Incentives for Green Tech Adoption

Government incentives are critical in accelerating the adoption of green technologies in transportation, such as electric vehicles (EVs), hydrogen fuel systems, and smart infrastructure. These technologies, while promising in their environmental benefits, often entail high initial costs and require robust support systems to become commercially viable and widely adopted.

Fiscal incentives—including tax credits, subsidies, and grants—are among the most effective tools governments use to support both producers and consumers of sustainable transportation technologies. Countries like Norway and the Netherlands have achieved high EV penetration rates by exempting EVs from import taxes, offering free public parking, and allowing access to high-occupancy vehicle lanes (Figenbaum, 2017). Such policies have made EVs more attractive to consumers and signaled strong governmental commitment to sustainability goals.

Research and development (R&D) funds and public-private partnerships are two ways in which governments may encourage innovation in industry. Advancements in electric drivetrain technologies, charging infrastructure, and battery technology are supported by the Vehicle Technologies Office of the United States Department of Energy (DOE, 2021). Sustainable mobility solutions have been commercialised more quickly because of creative ecosystems that have been established via similar projects in Asia and Europe.

Incentives must also be equitable and targeted. Overly broad subsidies can disproportionately benefit higher-income populations who can afford premium EVs, thereby reinforcing inequality (Jenn, 2019). Designing inclusive policies—such as offering higher rebates for low-income households or funding public EV fleets—can ensure a fair distribution of green technology benefits.

Regulatory mandates like emissions targets and zero-emission vehicle quotas can complement financial incentives by setting clear industry expectations. These mandates push automotive manufacturers and logistics providers to innovate and invest in cleaner alternatives, creating market momentum for sustainable transport.

6.3 Need for International Standards and Ethical Frameworks

The global nature of transportation systems and supply chains necessitates the development of international standards and ethical frameworks to govern the use of emerging technologies. Autonomous vehicles, AI-driven logistics, and drone-based deliveries operate across borders and jurisdictions, making consistent regulatory oversight essential to ensure safety, interoperability, and fairness.

One critical area is the development of safety and operational standards for autonomous vehicles. International organizations such as the International Organization for Standardization (ISO) and the United Nations Economic Commission for Europe (UNECE) are actively working on harmonizing protocols for AV testing, data sharing, and human-machine interaction (UNECE, 2020). These standards aim to create a shared framework that enables countries to adopt AVs while minimizing regulatory fragmentation.

Ethical considerations are equally important, particularly with respect to algorithmic decision-making and data governance. AI systems used in transport planning or autonomous navigation may encode biases or make decisions with significant social implications. Ethical frameworks should guide developers and policymakers in ensuring transparency, accountability, and fairness in algorithmic systems (Jobin et al., 2019). This includes ensuring that data collected from passengers or workers is used responsibly and in compliance with privacy laws like the General Data Protection Regulation (GDPR) in Europe.

Global cooperation is needed to manage the environmental externalities of green technologies, such as lithium mining for batteries or electronic waste from outdated transport systems. International treaties and environmental impact standards can support the sustainable sourcing and disposal of technological components, ensuring that the green transition does not create new ecological harm.

Multilateral forums like the United Nations, G20, and the World Bank can facilitate dialogue and cooperation among nations, enabling knowledge-sharing and coordinated policy responses to technological disruption in transport. This collaborative approach is essential to ensure that the benefits of innovation are equitably shared across developed and developing countries.

The transition to a technology-driven transportation sector brings with it immense opportunities for sustainability and efficiency but also presents regulatory and ethical challenges. Labor protections must adapt to safeguard workers in an automated economy, while targeted government incentives can accelerate the equitable adoption of green technologies. Furthermore, the establishment of international standards and ethical frameworks is crucial for ensuring global alignment in safety, sustainability, and fairness. As technological innovation continues to redefine the future of transport, responsive and inclusive policy-making will be essential to realizing its full potential.

7. STRATEGIC RECOMMENDATIONS

As the transportation sector undergoes rapid transformation driven by emerging technologies such as automation, electrification, and artificial intelligence, a multi-stakeholder strategy is essential to maximize benefits while mitigating potential adverse impacts. Governments, industry stakeholders, and educational institutions each play a critical role in shaping a sustainable, inclusive, and forward-looking transport ecosystem. This section outlines strategic recommendations for each stakeholder group to promote digital readiness, workforce inclusivity, and technological competency.

7.1 For Governments: Investment in Digital Infrastructure and Retraining Programs

Governments must act as enablers of change by investing in both hard and soft infrastructure to support the evolving transportation landscape. One of the most pressing needs is the expansion and modernization of digital infrastructure to facilitate the deployment of smart transport technologies. High-speed internet, 5G networks, and data-sharing platforms are foundational to the operation of autonomous vehicles (AVs), Internet of Things (IoT) systems, and AI-based traffic management (OECD, 2020). Without adequate digital infrastructure, the full benefits of technology-driven mobility solutions cannot be realized.

Governments should establish robust retraining and upskilling programs to mitigate the risk of widespread job displacement. As automation and AI technologies reduce the demand for traditional roles such as drivers, dispatchers, and mechanics, reskilling initiatives can help displaced workers transition into emerging roles, such as AV maintenance technicians, cybersecurity specialists, and data analysts (Subrahmanyam, 2025). These programs should be inclusive and accessible, targeting vulnerable populations, including low-income individuals and those with limited digital literacy.

Public investment in workforce development can be modeled after successful examples such as Singapore's SkillsFuture initiative, which provides citizens with credits to pursue lifelong learning in high-demand fields, including transport technologies (SkillsFuture SG, 2021). In addition to direct funding, governments should incentivize private-sector partnerships and apprenticeships to ensure alignment between training content and labor market needs.

Policy coherence is also essential. Governments should integrate workforce development strategies into broader transport and sustainability policies, ensuring that employment transitions are planned alongside technology deployments. By adopting a whole-of-government approach, policymakers can support economic resilience while advancing technological innovation.

7.2 For Industry Stakeholders: Inclusive Employment Models and Innovation-Driven Hiring

Industry stakeholders—including automotive manufacturers, logistics providers, and ride-hailing platforms—must adopt proactive labor strategies to create a future-ready workforce. One key recommendation is the adoption of inclusive employment models that ensure fair treatment, social protection, and equitable access to opportunities. As non-standard employment (e.g., gig work, contract labor) becomes more prevalent in the digital transport economy, companies must revisit compensation structures, benefits, and worker protections to foster long-term engagement and stability (ILO, 2021).

Inclusive models should also promote diversity in hiring, particularly in technological roles traditionally dominated by specific demographics. Recruitment practices must be reoriented to value adaptability, learning potential, and transferable skills, rather than only formal qualifications. Hiring based on these broader competencies allows for greater participation from underrepresented groups, including women, minorities, and workers without formal STEM education (World Economic Forum, 2020).

In parallel, industry must embrace innovation-driven talent acquisition. As transport systems become increasingly digitalized, businesses should build interdisciplinary teams capable of integrating AI, robotics, software engineering, and systems design. This requires not only hiring new talent but also creating internal career mobility pathways to redeploy existing employees into tech-focused roles.

Collaboration with start-ups, academic institutions, and public agencies can further enhance innovation capacity. For example, companies can invest in innovative labs and centers of excellence that incubate new ideas and train employees in emerging technologies. The goal should be to create an organizational culture that values continuous learning and technological experimentation.

Finally, responsible corporate behavior is paramount. Industry leaders should contribute to the development of ethical standards and best practices around data privacy, algorithmic fairness, and sustainability in transportation technologies. Doing so not only builds trust with consumers and regulators but also enhances brand reputation in an increasingly value-driven market.

7.3 For Educational Institutions: Curriculum Updates and Vocational Programs in Transport Tech

Educational institutions, from universities to vocational training centers, are at the forefront of preparing the workforce for future transportation roles. They must respond to evolving industry demands by updating curricula, developing interdisciplinary programs, and fostering practical skills aligned with transport technology trends.

Curriculum updates should incorporate key areas such as artificial intelligence, machine learning, battery and electric vehicle systems, autonomous navigation, and transport logistics. These technical components must be supplemented with courses on ethics, sustainability, and regulatory compliance to ensure holistic education. Universities should also embed real-world projects, simulations, and internships into programs to provide experiential learning opportunities (OECD, 2019).

Vocational training plays an equally vital role, particularly for workers transitioning from declining sectors. Institutions should design short-term certification and diploma programs tailored to adult learners, focusing on hands-on competencies like EV maintenance, drone operation, and digital fleet management. Collaborations with industry can ensure that training content is current and relevant, increasing graduates' employability.

Educational institutions should foster partnerships with government and industry to co-create innovation ecosystems. Initiatives like Germany's dual-education system and Canada's Mitacs program, which links students with companies through research internships, offer replicable models (Bahl & Dietzen, 2020). These partnerships can also support lifelong learning through modular, stackable credentials that accommodate diverse learning paths and career trajectories.

Institutions must invest in teacher and faculty development. As the demand for instruction in advanced technologies grows, educators must be equipped with both technical knowledge and pedagogical tools to deliver high-quality learning experiences.

Finally, educational access must be equitable. Scholarships, outreach programs, and flexible learning formats can enable underprivileged students to participate in transport tech education, contributing to a more inclusive digital economy.

The transformation of the transportation sector requires coordinated efforts across governments, industry stakeholders, and educational institutions. Governments must lead with infrastructure investments and worker transition strategies, while industry stakeholders must foster inclusive, future-oriented employment models. Educational institutions, in turn, must evolve their curricula and delivery models to meet the demands of a technology-driven transport economy. By working collaboratively, these stakeholders can ensure that the future of transportation is not only efficient and innovative but also inclusive, equitable, and sustainable.

8. FUTURE OUTLOOK

Quick technological developments, changing social expectations, and pressing climate imperatives are causing a sea change in the transport industry. New service models like Mobility-as-a-Service (MaaS), complete automation of logistics and mobility systems, and the mixing of human and artificial intelligence (AI) in workforce design will determine the future of transportation. In this part, we will look at how MaaS jobs are changing, what the future holds for automation and sustainability objectives, and how scenario planning might help society be ready for a workforce that combines human and AI workers.

8.1 Emerging Roles in Mobility-as-a-Service (MaaS)

The advent of mobility-as-a-service, or MaaS, has marked a sea change away from individual car ownership and towards unified, customer-focused transportation options. MaaS solutions integrate many modes of transportation into one user-friendly platform, including public transportation, ride-sharing, micro-mobility, and bike-sharing (Kamargianni et al., 2016). Jobs in the transport sector are proliferating as a result of this model's rising popularity throughout the world.

First, the MaaS sector is creating demand for data scientists, software developers, and UX/UI designers who can build and maintain seamless, interoperable mobility platforms. These professionals are responsible for developing routing algorithms, integrating diverse service providers, and ensuring cybersecurity within the platform. As user data becomes central to service personalization, privacy experts and compliance officers will also become essential.

Second, customer service and digital operations roles are evolving. Unlike traditional transport operators, MaaS providers often offer real-time customer support, requiring agile teams with strong digital and interpersonal skills. Mobility consultants and accessibility coordinators are emerging roles, helping cities and users navigate multimodal transport solutions while ensuring inclusivity and equity.

Transportation planners are being redefined. Rather than focusing solely on physical infrastructure, these professionals must now also design digital mobility networks and coordinate partnerships across public and private entities. Their tasks include forecasting demand, analyzing mobility patterns, and developing MaaS governance structures (Mulley & Nelson, 2020).

Thus, MaaS introduces a hybrid employment model—one that merges the tech sector's innovation dynamics with the service orientation of the traditional transport industry. This demands adaptable, tech-savvy workers prepared to engage with evolving digital ecosystems.

8.2 Long-Term Impacts of Full Automation and Climate Goals

The long-term trajectory of the transportation sector is shaped by two powerful forces: automation and climate change mitigation. As AVs, drones, and automated logistics systems become mainstream, the sector will experience fundamental reconfiguration in job roles and environmental impact.

Full automation in transport could significantly reduce the need for human drivers, warehouse workers, and fleet dispatchers. According to a report by the International Transport Forum (ITF, 2017), up to 70% of truck-driving jobs in some countries could be automated by 2030. This potential displacement necessitates large-scale workforce transition strategies, including reskilling and labor reallocation to emerging sectors.

The decarbonization of transport—central to achieving the goals of the Paris Agreement—will demand green innovation. The shift toward electric vehicles (EVs), hydrogen-powered transport, and low-emission public transit will create new roles in battery manufacturing, renewable energy integration, and sustainable logistics (IEA, 2022). The rise of circular economy practices in EV production will spur demand for specialists in battery recycling and material recovery.

Long-term employment in transport will likely shift from operational to supervisory and technical domains. Human oversight will remain critical in safety-critical systems, ethical AI governance, and climate impact monitoring. Infrastructure development will pivot toward green, smart, and adaptive systems, requiring engineers and urban planners with advanced environmental and technological knowledge.

The intersection of full automation and climate goals could also foster synergies—such as energy-efficient autonomous transport or AI-optimized route planning to minimize emissions. Without proactive governance, these technologies may exacerbate inequalities or lead to rebound effects (e.g., increased transport demand offsetting efficiency gains). This necessitates anticipatory policy and inclusive planning frameworks.

8.3 Scenario Planning for a Hybrid Human-AI Workforce

To navigate this complex future, scenario planning has emerged as a vital tool for policymakers, industry leaders, and educators. This method helps stakeholders envision multiple plausible futures and design flexible, adaptive strategies accordingly (Schoemaker, 1995).

One key scenario involves the rise of a hybrid human-AI workforce. In this future, AI complements human capabilities rather than replacing them outright. Autonomous trucks might still require remote human operators for supervision in complex environments. AI-based fleet management systems will depend on human analysts to interpret trends, optimize performance, and respond to anomalies. In such contexts, human roles shift toward oversight, critical thinking, and ethical judgment—skills that machines are unlikely to master fully.

Another scenario envisions widespread task reconfiguration. Jobs will not disappear entirely but will be unbundled and redistributed. A driver might evolve into a mobility ambassador who manages autonomous vehicles, provides human interaction during journeys, and troubleshoots digital systems. Similarly, logistics workers might transition into drone operations managers or inventory data analysts.

Scenario planning also reveals challenges. AI systems, if not properly designed and regulated, could reinforce biases, exclude marginalized populations, or undermine labor rights. Thus, ethical foresight, inclusive design, and interdisciplinary collaboration will be essential to guide technological integration responsibly.

Education systems must respond to these scenarios by promoting meta-skills—such as adaptability, problem-solving, and emotional intelligence—alongside digital and STEM competencies. Meanwhile, labor policies must support portable benefits, gig worker protections, and transition assistance to ensure that all workers can participate in the new mobility economy.

The future of the transportation sector lies at the intersection of innovation, sustainability, and inclusiveness. Emerging roles in MaaS signal a shift toward integrated, digitally mediated mobility systems. The twin forces of full automation and climate action will reshape employment structures and require coordinated transitions. Scenario planning provides a strategic lens to prepare for diverse futures, especially the emergence of hybrid human-AI work models. Stakeholders across sectors must collaborate to ensure that the future of transport is not only technologically advanced but also socially just and environmentally sustainable.

9. CONCLUSION

The transformation of the transportation sector illustrates the intricate interplay between technological innovation, sustainability imperatives, and evolving employment landscapes. As emerging technologies—ranging from autonomous vehicles and electric mobility to AI-powered logistics and smart infrastructure—reshape urban mobility, they simultaneously disrupt traditional roles while creating new, tech-oriented opportunities. These changes are not occurring in isolation; they are embedded within broader societal goals such as decarbonization, digital inclusion, and equitable economic growth. Hence, transport innovation must be understood not merely as a technological shift but as a socio-economic evolution with profound implications for cities, labor markets, and the environment.

A key insight from this examination is the necessity of proactive planning to manage the workforce transition. The potential displacement of low- and mid-skilled transport workers calls for deliberate investments in reskilling, upskilling, and the creation of accessible educational pathways. As digital and green competencies become foundational in the new mobility economy, the alignment of industry demands with academic curricula and vocational training becomes imperative. Governments, industries, and educational institutions must not only anticipate changes but also lead the way in shaping a resilient, inclusive workforce.

Finally, achieving sustainable and equitable urban mobility demands collaborative frameworks. Public-private partnerships, international regulatory coordination, and inclusive policy-making are essential to ensure that technological advancements benefit all strata of society. Special attention must be paid to marginalized communities to avoid deepening socio-economic divides through uneven access to mobility and employment. By fostering shared responsibility and leveraging collective intelligence, stakeholders can guide the transportation sector toward a future that is environmentally sound, economically inclusive, and socially just. The convergence of technology, sustainability, and employment should thus serve as a cornerstone for building smarter, greener, and fairer urban futures.

REFERENCES

Bahl, A., & Dietzen, A. (2020). *Work-based learning as a driver for skills development and job creation*. International Labour Organization.

Batarseh, F. A., & Freeman, L. A. (2021). *AI and machine learning for transport planning and operations*. Elsevier.

Batty, M., Axhausen, K. W., Giannotti, F., Pozdnoukhov, A., Bazzani, A., Wachowicz, M., Ouzounis, G., & Portugali, Y. (2012). Smart cities of the future. *The European Physical Journal. Special Topics, 214*(1), 481–518. DOI: 10.1140/epjst/e2012-01703-3

Bureau of Labor Statistics. (2023). *Occupational outlook for heavy and tractor-trailer truck drivers*.

Chambers, J., Barton, J., & Elser, M. (2020). *The circular economy and green jobs in transport*. International Transport Forum.

Cugurullo, F. (2020). Urban artificial intelligence: From automation to autonomy in the smart city. *Frontiers in Sustainable Cities, 2*, 38. DOI: 10.3389/frsc.2020.00038

De Stefano, V. (2016). The rise of the "just-in-time workforce": On-demand work, crowdwork, and labor protection in the "gig-economy.". *Comparative Labor Law & Policy Journal, 37*(3), 471–504.

Department of Energy (DOE). (2021). Vehicle Technologies Office.

European Commission. (2021). *Fit for 55: Delivering the EU's 2030 climate target on the way to climate neutrality*.

Figenbaum, E. (2017). Perspectives on Norway's supercharged electric vehicle policy. *Environmental Innovation and Societal Transitions, 25*, 14–34. DOI: 10.1016/j.eist.2016.11.002

Frey, C. B., & Osborne, M. A. (2017). The future of employment: How susceptible are jobs to computerisation? *Technological Forecasting and Social Change, 114*, 254–280. DOI: 10.1016/j.techfore.2016.08.019

Ghosh, D., & Choi, J. P. (2022). Cybersecurity challenges in smart transportation systems. *Transportation Research Part C, Emerging Technologies, 140*, 103719.

Gittleman, M., & Monaco, K. (2020). *Truck automation and job displacement in the United States*. University of Pennsylvania.

Hilmola, O. P., & Henttu, V. (2022). Transformation of logistics industry towards environmental sustainability and automation. *Sustainability, 14*(1), 118.

Hussain, R., Son, J., Eun, H., & Oh, H. (2020). Internet of Vehicles: Challenges and future technologies. *IEEE Access : Practical Innovations, Open Solutions, 8*, 146173–146197.

IEA. (2023). *Global EV Outlook 2023: Catching up with climate ambitions*. International Energy Agency.

ILO. (2021). *World employment and social outlook: The role of digital labour platforms in transforming the world of work*. International Labour Organization.

CB Insights. (2022). *Autonomous trucks: 40+ companies working on driverless trucking*.

SAE International. (2021). *Taxonomy and Definitions for Terms Related to Driving Automation Systems for On-Road Motor Vehicles* (J3016_202104).

International Energy Agency (IEA). (2022). *Global EV Outlook 2022*.

International Energy Agency (IEA). (2022).. . *Transport*.

International Energy Agency (IEA). (2023). *Global EV Outlook 2023*.

International Labour Organization (ILO). (2018). [*Greening with jobs.*]. *World Employment and Social Outlook*, •••, 2018.

International Labour Organization (ILO). (2021). [*The role of digital labour platforms in transforming the world of work.*]. *World Employment and Social Outlook*, •••, 2021.

International Transport Forum (ITF). (2017). *Managing the Transition to Driverless Road Freight Transport*.

Jenn, A. (2019). Equity impacts of electric vehicle policies in the United States. *Transportation Research Part D, Transport and Environment, 67*, 1–8.

Jobin, A., Ienca, M., & Vayena, E. (2019). The global landscape of AI ethics guidelines. *Nature Machine Intelligence, 1*(9), 389–399. DOI: 10.1038/s42256-019-0088-2

Kamargianni, M., Matyas, M., Li, W., & Muscat, J. (2016). *Londoners' attitudes towards car-ownership and Mobility-as-a-Service: Impact assessment and opportunities that lie ahead*. UCL Energy Institute.

Lucas, K. (2019). Transport and social exclusion: Where are we now? *Transport Policy, 20*, 105–113. DOI: 10.1016/j.tranpol.2012.01.013

Manyika, J., Lund, S., Chui, M., Bughin, J., Woetzel, J., Batra, P., & Sanghvi, S. (2017). *Jobs lost, jobs gained: Workforce transitions in a time of automation.* McKinsey Global Institute.

Martens, K. (2017). *Transport justice: Designing fair transportation systems.* Routledge.

Maurya, S., Shavkidinova, D., Shrivastava, V., & Subrahmanyam, S. (2025). Effects of Green Supply Chain Management Practices on Environmental and Firm Sustainability Performance. *ACTA INNOVATIONS*, *55*, 13–22.

McKinsey & Company. (2022). *The future of work in transportation: Preparing for technological disruption.* https://www.mckinsey.com/industries/travel-logistics-and-infrastructure/our-insights/the-future-of-work-in-transportation

McKinsey & Company. (2023). *The future of automation in logistics: Robotics and drones.*

Milakis, D., van Arem, B., & van Wee, B. (2017). Policy and society related implications of automated driving: A review of literature and directions for future research. *Journal of Intelligent Transport Systems*, *21*(4), 324–348. DOI: 10.1080/15472450.2017.1291351

Ministry of Manpower Singapore. (2021). *Strengthening protections for platform workers.*

Mohamed, M., Ferguson, M., & Ahmad, M. (2020). The role of artificial intelligence in enhancing public transport systems. *Transportation Research Interdisciplinary Perspectives*, *4*, 100091.

Mulley, C., & Nelson, J. D. (2020). Mobility as a Service (MaaS) – does it have critical mass? *Transport Reviews*, *40*(3), 239–243.

OECD. (2019). *Future of Education and Skills 2030: Curriculum analysis.* Organisation for Economic Co-operation and Development.

OECD. (2020). *The digital transformation of SMEs.*

OECD. (2022). *Skills for a 21st century transport workforce.* Organisation for Economic Co-operation and Development.

OECD. (2022). *Skills for the digital transformation of logistics and transport.*

Otto, A., Agatz, N., Campbell, J., Golden, B., & Pesch, E. (2018). Optimization approaches for civil applications of unmanned aerial vehicles (UAVs) or aerial drones: A survey. *Networks*, *72*(4), 411–458. DOI: 10.1002/net.21818

Pojani, D., & Stead, D. (2017). *The urban transport crisis in emerging economies*. Springer. DOI: 10.1007/978-3-319-43851-1

Rosenblat, A., & Stark, L. (2021). Algorithmic labor and information asymmetries: A case study of Uber's drivers. *International Journal of Communication*, *10*, 3758–3784.

Schoemaker, P. J. H. (1995). Scenario planning: A tool for strategic thinking. *Sloan Management Review*, *36*(2), 25–40.

Sharma, R., & Ghosh, D. (2021). Smart urban transport: Predictive analytics for transit management. *International Journal of Transportation Science and Technology*, *10*(1), 1–11.

Shaw, C., Hales, J., & Li, B. (2020). Gender and urban mobility: Perspectives from low-income communities. *Journal of Transport Geography*, *87*, 102813.

SkillsFuture SG. (2021). *About SkillsFuture*.

Smith, A., & Anderson, J. (2021). *AI and the future of transportation jobs*. Pew Research Center.

Subrahmanyam, S. (2025). Data Governance and Privacy in Smart Cities. In Sharma, A., Mansotra, V., & Singh, P. (Eds.), *Citizen-Centric Artificial Intelligence for Smart Cities* (pp. 433–456). IGI Global Scientific Publishing. DOI: 10.4018/979-8-3693-7832-8.ch017

Subrahmanyam, S. (2025). Employee Skillsets in the Digital Age. In Özbek, M., & Christiansen, B. (Eds.), *Organizational Sociology in the Digital Age* (pp. 369–388). IGI Global Scientific Publishing. DOI: 10.4018/979-8-3693-7398-9.ch018

Subrahmanyam, S. (2025). Role of Artificial Intelligence in Re-Inventing Human Resource Management (HRM) 2.0. In Kapoor, T., Singh, A., Sahay, K., Kaur, H., & Verma, R. (Eds.), *Sustainable Management Practices for Employee Retention and Recruitment* (pp. 243–264). IGI Global Scientific Publishing. DOI: 10.4018/979-8-3693-8679-8.ch012

Subrahmanyam, S., Azoury, N., & Sarkis, N. (2024). Optimising Energy Efficiency through Effective Supply Chain Management: A Comparative Study. *ACTA INNOVATIONS*, *51*(1), 46–51. DOI: 10.62441/ActaInnovations.51.7

Subrahmanyam, S., Khalife, D., & El-Chaarani, H. (2024). Towards Sustainable Future: Exploring Renewable Energy Solutions and Environmental Impacts. *ACTA INNOVATIONS*, *51*(1), 15–24. DOI: 10.62441/ActaInnovations.51.3

Udacity. (2021). *Mercedes-Benz and Udacity: Collaborating on autonomous systems education*.

United Nations Economic Commission for Europe (UNECE). (2020). *Framework for automated/autonomous vehicles.*

USDOT. (2022). *Transportation workforce development and training programs.* U.S. Department of Transportation.

Weiss, M., Cloos, M., & Kasten, P. (2020). *The impact of electrification on the German automotive workforce*. Fraunhofer Institute.

World Bank. (2022). *Transforming transportation: Smart mobility for sustainable cities.*

World Economic Forum. (2020). *The future of jobs report 2020.*

Wurman, P. R., D'Andrea, R., & Mountz, M. (2008). Coordinating hundreds of cooperative, autonomous vehicles in warehouses. *AI Magazine*, *29*(1), 9–20.

Zhao, Y., Gao, H., Zhang, L., & Wang, Y. (2019). Intelligent transportation systems: When AI meets transportation. *Transportation Research Part C, Emerging Technologies*, *100*, 1–4. DOI: 10.1016/j.trc.2019.01.013

Chapter 7
Emerging Technologies in Multi-Modal Networks: Graph Theory Solutions

Suvarna Mohanrao Nade
https://orcid.org/0009-0001-7538-899X
Sandip University, Nashik, India

Renu Praveen Pathak
https://orcid.org/0000-0001-5310-3722
Sandip University, Nashik, India

ABSTRACT

Multi-modal transportation networks are critical to modern logistics and urban mobility, requiring seamless integration across diverse transportation modes. Emerging technologies, such as artificial intelligence (AI), the Internet of Things (IoT), and big data, offer transformative opportunities for optimizing these networks. Graph theory, a robust mathematical framework, is a powerful tool for addressing complex challenges like route optimization, congestion management, and resource allocation in multi-modal systems. This chapter explores the confluence of these technologies and graph theory, highlighting innovative solutions for enhancing efficiency, sustainability, and resilience in transportation networks. Case studies and a synthesis of recent research underscore the role of graph-based algorithms in enabling adaptive and intelligent transportation systems while identifying future trends and research opportunities.

DOI: 10.4018/979-8-3373-0882-1.ch007

INTRODUCTION

With the rapid advancements in transportation systems, multi-modal networks have become increasingly complex, requiring innovative methodologies for their optimization and management. Multi-modal transportation networks integrate various modes of transport, such as road, rail, air, and maritime, to enhance connectivity and efficiency. The study and enhancement of these networks demand sophisticated analytical tools, among which graph theory plays a pivotal role. Emerging technologies such as Artificial Intelligence (AI), the Internet of Things (IoT), and blockchain have further transformed the scope of multi-modal network analysis, making real-time data processing and decision-making more efficient (Zhang, Liu, & Chen, 2021).

Graph theory provides a mathematical framework to model, analyze, and optimize transportation networks by representing them as graphs consisting of vertices (nodes) and edges (connections). This framework aids in route optimization, congestion management, and network resilience assessment. As cities and transportation infrastructures expand, the integration of graph theory with emerging technologies has the potential to revolutionize the efficiency and sustainability of multi-modal networks (Bertsimas & Tsitsiklis, 2020).

Multi-Modal Networks: Definition, structure and Importance

A multi-modal transportation network comprises different transportation systems interconnected to allow seamless movement of passengers and goods. These networks are crucial for modern urban mobility and logistics, aiming to minimize travel time, reduce congestion, and lower carbon footprints (Hansen, Huang, & Wang, 2022). A multimodal network consists of nodes and edges, where nodes represent locations (e.g., airports, railway stations, bus stops), and edges depict connections between them. These connections may involve different transport modes, requiring the optimization of routes and transfer points to minimize travel time and cost (Boccaletti et al., 2014). The integration of various transport modes, however, introduces complexity in network planning and management. This complexity necessitates advanced analytical approaches such as graph theory to optimize routes, minimize costs, and enhance service reliability.

Significance of Multimodal Networks

1. **Enhanced Connectivity**: Multimodal networks allow seamless movement across different transportation systems, reducing travel time and congestion. For instance, integrating railway and air travel improves passenger transfer efficiency at major hubs.

2. **Resilience and Redundancy**: These networks enhance resilience by providing alternative routes and transport modes in case of disruptions, such as road closures or flight delays (Zhang, Zhang, & Chen, 2020).
3. **Optimization of Resources**: Multimodal transportation reduces reliance on a single mode, leading to cost savings and improved resource utilization, particularly in freight logistics.
4. **Sustainability**: By integrating eco-friendly transportation options, such as public transit and cycling, multimodal networks contribute to reduced carbon emissions and sustainable urban development.
5. **Applications in Air Transportation**: In aviation, multimodal networks facilitate better airport access, improving passenger and cargo transport efficiency through integrated rail-air or road-air connectivity. Graph theory-based approaches are increasingly applied to analyze and optimize these networks for enhanced operational efficiency (Guimerà, Mossa, Turtschi, & Amaral, 2005).

The Role of Graph Theory in Multi-Modal Networks

Graph theory is a fundamental mathematical tool used to model multi-modal networks efficiently. By representing a network as a graph, researchers and planners can analyze various aspects, such as:

- **Shortest Path Algorithms**: Algorithms like Dijkstra's and A* help determine the most efficient routes across different modes of transport (Cormen, Leiserson, Rivest, & Stein, 2009).
- **Network Resilience Analysis**: Identifying critical nodes and links to ensure robustness against disruptions (Wang, Xu, & Zhou, 2023)(Wang, Yu, & Zhao, 2023)(Wang, Zhou, & Feng, 2023).
- **Traffic Flow Optimization**: Using flow algorithms to manage congestion and improve service efficiency (Bell & Iida, 2019).
- **Connectivity and Accessibility Measures**: Evaluating network connectivity to ensure equitable access to transportation services (Newman, 2018).

EMERGING TECHNOLOGIES ENHANCING GRAPH THEORY APPLICATIONS

Several emerging technologies are significantly augmenting the application of graph theory in multi-modal networks:

1. Artificial Intelligence (AI) and Machine Learning

AI-driven approaches improve predictive analytics in transportation networks. Machine learning models trained on historical and real-time data can predict traffic patterns, optimize scheduling, and enhance route planning (Zhao, Wang, & Sun, 2022) (Zhao, He, & Chen, 2022). Deep learning techniques, when combined with graph neural networks (GNNs), provide insights into complex multi-modal interactions.

Figure 1. flowchart

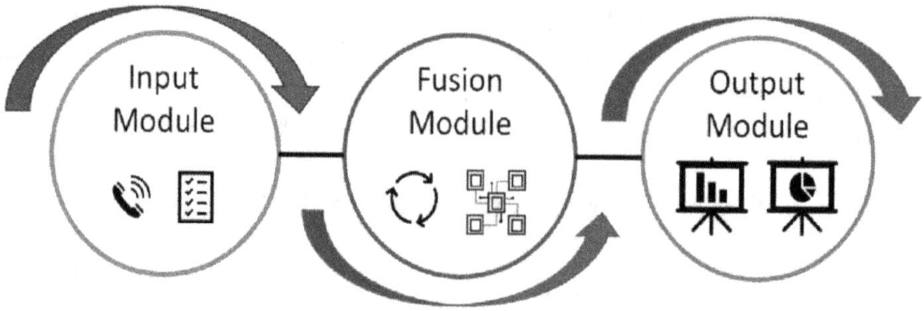

Figure 1 illustrates a high-level architecture of an AI and Machine Learning system. The Input Module processes diverse data types such as text, audio, and video, enabling multimodal learning. The Fusion Module integrates these heterogeneous data sources to extract meaningful patterns and insights using AI algorithms and machine learning models. The Output Module generates visual or analytical results for interpretation and decision-making. This pipeline showcases how AI systems process raw inputs, analyze them through fusion techniques, and deliver structured outputs. Such models are essential in applications like speech recognition, multimedia analysis, and intelligent decision support systems.

2. Internet of Things (IoT)

IoT-enabled sensors and connected devices provide real-time data on traffic flow, vehicle locations, and infrastructure conditions. Graph-based algorithms use this data to dynamically adjust transportation routes and schedules, improving overall efficiency (Perera, Liu, & Jayawardena, 2017).

Figure 2. the field of the internet of things

Figure 2 highlights key application areas of the Internet of Things (IoT) across various domains. IoT technology connects physical devices through the internet, enabling real-time data exchange and automation. As illustrated, it spans multiple fields such as Traffic (smart vehicles), Industry (automation and monitoring), Social (communication and networking), Medical (remote healthcare and monitoring), Home (smart home devices), Power Grid (smart energy management), and City (smart city infrastructure). These interconnected systems enhance efficiency, safety, and convenience, showcasing the transformative potential of IoT in modern life and its crucial role in building intelligent, responsive environments.

3. Blockchain Technology

Blockchain ensures transparency and security in transportation networks by facilitating secure data sharing among stakeholders. Smart contracts automate transactions in logistics and freight management, enhancing trust and efficiency (Saberi, Kouhizadeh, & Sarkis, 2019). Blockchain Technology in Multimodal Networks ensures secure, transparent, and decentralized data exchange among various transportation and communication modes. By recording transactions across distributed ledgers, blockchain enhances trust, traceability, and accountability in logistics, ticketing, and asset tracking. Smart contracts automate processes like payments and cargo transfers, reducing delays and fraud. In multimodal transport systems, it supports seamless coordination between stakeholders—rail, air, road, and sea—while maintaining data integrity. Blockchain also strengthens cybersecurity and privacy in data sharing, especially in real-time analytics. Its integration leads to more efficient, reliable, and user-centric multimodal networks, fostering innovation in mobility and infrastructure management.

4. Big Data Analytics

The vast amount of data generated from transportation networks requires advanced analytical tools for processing. Graph-based big data analytics enables pattern recognition, anomaly detection, and predictive maintenance, leading to more resilient network designs (Ghosh, Bose, & Mukherjee, 2021). Big Data Analytics in Multimodal Networks involves processing and analyzing vast volumes of heterogeneous data from various sources such as sensors, GPS, social media, and IoT devices. By integrating diverse data modalities—text, audio, video, and spatial—graph-based models enable pattern recognition, predictive insights, and anomaly detection. Big Data tools like Hadoop, Spark, and cloud platforms support real-time processing and scalable storage. In multimodal transportation networks, this enhances decision-making, route optimization, and user experience. With the rise of Graph Neural Networks (GNNs), big data analytics facilitates deeper understanding of dynamic relationships and fosters smarter, resilient, and more efficient networked systems.

5. Quantum Computing

Quantum computing holds the potential to solve complex optimization problems in multi-modal networks at unprecedented speeds. Graph-theoretic problems, such as the Traveling Salesman Problem (TSP) and Vehicle Routing Problem (VRP), can be efficiently solved using quantum algorithms, optimizing logistics and urban mobility (Arute et al., 2019).

The rapid evolution of multi-modal transportation networks has necessitated the integration of advanced computational techniques to enhance efficiency, reliability, and sustainability. Graph theory, with its robust mathematical framework, has emerged as a powerful tool for optimizing network connectivity, reducing congestion, and improving overall system performance (Ahuja, Magnanti, & Orlin, 1993) (Bell & Iida, 1997). With the increasing complexity of transportation systems, the adoption of emerging technologies such as artificial intelligence (AI), the Internet of Things (IoT), and real-time data analytics has further strengthened decision-making and operational efficiency (Chen, Xu, & Zhang, 2021) (Zhang, Wang, & Chen, 2022). This chapter explores how graph theory, in combination with these technologies, can address challenges in multi-modal transportation networks.

Multimodal networks are fundamental to modern transportation systems, enabling efficient, sustainable, and resilient mobility solutions. The application of graph theory and network science in analyzing these networks continues to enhance their functionality and adaptability in real-world scenarios.

KEY RESEARCH QUESTIONS

This chapter aims to address the following key research questions:

- How can graph theory be applied to optimize multi-modal transportation networks?
- What emerging technologies are enhancing the efficiency and reliability of multi-modal networks?
- How do graph-based algorithms contribute to route optimization, congestion management, and network resilience?
- What are the challenges and limitations of integrating graph theory with modern transportation technologies?
- How can real-time data and AI-driven approaches be leveraged within graph-theoretic frameworks for network optimization?
- What are the future trends and research opportunities in applying graph theory to multi-modal transportation?

Overview of the Chapter's Structure

This chapter is structured as follows:

Section 2 presents fundamental graph theory concepts and their applications in transportation networks, with a focus on classical and modern approaches (Dijkstra, 1959) (Ford & Fulkerson, 1962).
Section 3 explores emerging technologies that enhance multi-modal network performance, including AI, IoT, and big data analytics (Gkiotsalitis & Cats, 2021a) (Li, Wang, & Zhang, 2020).
Section 4 discusses graph-theoretic solutions for network optimization, particularly in route planning, congestion management, and resilience assessment (Lou, Sun, & Zhang, 2023) (Yuan, Liu, & Yang, 2019).
Section 5 examines key challenges and limitations associated with integrating graph-based models with modern transportation technologies.
Section 6 highlights future research directions, including advancements in quantum computing and AI-driven predictive models (Wang, Xu, & Zhou, 2023)(Wang, Yu, & Zhao, 2023)(Wang, Zhou, & Feng, 2023).

Finally, **Section 7** summarizes key findings and implications for researchers and practitioners.

This structured approach provides a comprehensive understanding of how graph theory, combined with emerging technologies, can revolutionize multi-modal transportation networks.

SECTION 2: FUNDAMENTALS OF MULTI-MODAL NETWORKS & GRAPH THEORY

Multi-modal networks integrate various modes of transportation or communication, forming interconnected systems that enhance efficiency and adaptability. Graph theory, a mathematical framework used to model and analyze such networks, plays a crucial role in optimizing their design, functionality, and resilience. This paper explores the fundamentals of multi-modal networks, their significance, and the role of graph theory in their study and application.

2. Understanding Multi-Modal Networks

2.1 Definition and Characteristics

A multi-modal network is a system comprising multiple interconnected modes of transport or communication, such as road, rail, air, and waterways or A multimodal network is a transportation system that integrates multiple modes of transport, such as road, rail, air, and waterways, to facilitate seamless movement of passengers or goods through efficient interconnections and transfer points (Rodrigue, 2020). These networks aim to optimize travel efficiency, reduce costs, and enhance accessibility by leveraging the advantages of different transport modes (Gkiotsalitis & Cats, 2021a) (Gkiotsalitis & Cats, 2021b). These networks are characterized by:

Integration of Multiple Transport Modes

Multimodal networks incorporate different modes of transportation, such as road, rail, air, and waterways, to provide efficient connectivity (Rodrigue, 2020).

Interconnectivity and Transferability

These networks facilitate seamless transfers between modes, reducing travel time and improving accessibility (Gkiotsalitis & Cats, 2021a) (Gkiotsalitis & Cats, 2021b).

Optimization for Efficiency

Multimodal networks are designed to minimize costs, travel time, and environmental impact by leveraging the strengths of different transport modes (Daganzo, 2019).

Complex Network Topology

The structure of multimodal networks often follows a graph-theoretical approach, representing nodes (stations, hubs) and edges (routes, links) with varying weights (Wu, Liu, & Zhu, 2020).

Reliability and Resilience

These networks enhance reliability by offering alternative routes and transport modes, reducing the impact of disruptions (Cats & Jenelius, 2018).

Sustainability and Environmental Considerations

By integrating sustainable transport options like public transit, cycling, and walking, multimodal networks contribute to reducing carbon footprints (Banister, 2019).

Smart Technology Integration

The use of intelligent transportation systems (ITS), real-time tracking, and digital payment systems enhances the efficiency of multimodal networks (Zhang, Liu, & Chen, 2021).

2.2 Components of Multi-Modal Networks

Nodes (Transfer Points & Terminals)

Nodes serve as critical points where different transportation modes intersect, such as airports, railway stations, bus terminals, and ports. These facilitate seamless transfers between modes, enhancing connectivity and reducing travel time. Efficient node design improves passenger and freight movement.

Links (Transport Routes & Connections)

Links shows the pathways connecting nodes, including roads, railways, air routes, and waterways. These connections determine travel efficiency, network coverage, and accessibility. Well-integrated links or connections ensure smooth transitions between modes, minimizing delays and congestion.

Figure 3. characteristics of multimodal networks

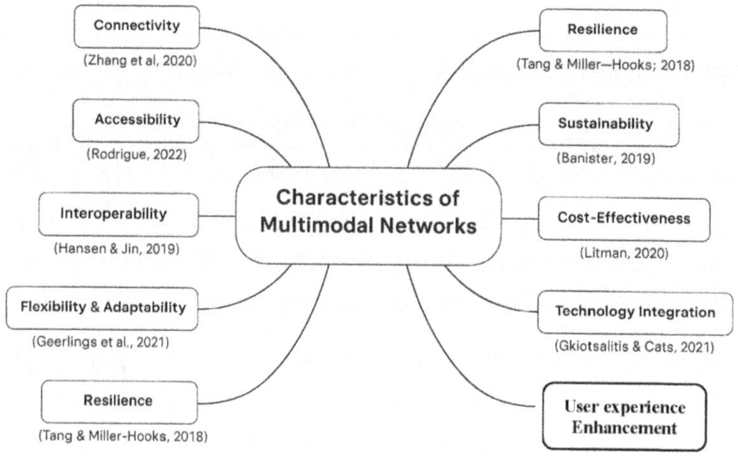

Figure 3 presents a visual representation of the key Characteristics of Multimodal Networks, emphasizing factors that contribute to enhanced user experience. These networks combine various transportation modes and infrastructures, requiring specific attributes to ensure efficiency, resilience, and usability. On the left, core operational and structural features are outlined: Connectivity (Zhang et al., 2020), Accessibility (Rodrigue, 2022), Interoperability (Hansen & Jin, 2019), and Flexibility & Adaptability (Geerlings et al., 2021). These attributes enable seamless transitions between transport modes and ensure inclusivity and mobility.

On the right, strategic and sustainability-focused characteristics are included: Resilience (Tang & Miller-Hooks, 2018), Sustainability (Banister, 2019), Cost-Effectiveness (Litman, 2020), and Technology Integration (Gkiotsalitis & Cats, 2021). These factors support long-term reliability, environmental responsibility, economic feasibility, and smart system integration.

Notably, Resilience appears on both sides, underscoring its dual importance in both system design and emergency responsiveness. The culmination of these interconnected features leads to User Experience Enhancement, which is the central goal of multimodal network development. By addressing these diverse yet interdependent characteristics, planners and engineers can build robust, user-centric transport systems that are adaptable to future challenges and evolving urban needs.

Modes of Transport

The network consists of different type transport modes, such as road (buses, cars), rail (trains, metros), air (flights), and waterways (ships, ferries). Each mode has unique advantages in terms of speed, cost, and capacity. The choice of mode depends on various factors like distance, infrastructure, and environmental considerations.

Intermodal Infrastructure

This includes bridges, tunnels, logistics hubs, and multimodal terminals that facilitate efficient movement between transport modes. High-quality infrastructure reduces bottlenecks, improves safety, and enhances operational efficiency. Smart infrastructure with digital tracking systems further optimizes multimodal transportation. The development of terminals, transfer hubs, and efficient logistics facilities enables smooth transitions between different transportation modes (Rodrigue, 2020).

Information & Communication Technology (ICT) Systems

Advanced ICT solutions, such as GPS tracking, real-time scheduling, and automated ticketing, enhance network efficiency. These technologies provide users with live updates on routes, delays, and alternative travel options. Digital platforms also improve freight tracking, reducing logistical uncertainties.

Regulatory & Policy Framework

Government policies, international agreements, and local regulations govern multimodal networks. Standardized policies ensure smooth coordination between transport modes, optimize freight movement, and promote sustainability. Regulatory frameworks also address safety, security, and environmental concerns.

Users & Demand Patterns

Passengers and freight operators form the core users of multimodal networks, each with specific needs and preferences. Travel demand varies based on factors such as urbanization, economic activity, and seasonal fluctuations. Understanding user behaviour helps in optimizing network design and service planning. In user experience enhancement Passenger-centric design, real-time tracking, and seamless transfers improve convenience and satisfaction (Smith & Brown, 2020).

2.2.1 Graph Theory concepts in Multi-Modal Networks

Graph theory provides the mathematical foundation for analyzing the structure and efficiency of multi-modal networks. A graph consists of a set of vertices (nodes) and edges (connections), where:

- **Nodes (Vertices, V)** – Represent entities in the network (e.g., airports in an air transportation network).
- **Edges (E)** – Represent connections between nodes (e.g. flight routes between airports).
- **Weights (w)** – Represent the cost, distance, or time associated with traveling along an edge.
- **Directed Graphs (Digraphs):** Represent asymmetric connections, such as one-way streets or airline routes.
- **Undirected Graphs:** Used in bidirectional systems, such as pedestrian pathways or railway tracks.
- **Multilayer Graphs:** Capture different modes in distinct layers but allow interactions between layers (Boccaletti et al., 2014).

2.2.2 Graph Types

Graph theory includes various types of graphs applicable to multi-modal networks:

- **Planar Graphs:** Graphs that can be drawn on a plane without edges crossing, often used in urban transportation layouts.
- **Non-Planar Graphs:** Networks where edges intersect, such as airline route maps.
- **Dynamic Graphs:** Change over time due to factors like traffic congestion or weather conditions.
- **Stochastic Graphs:** Involve probabilistic edge weights, representing uncertain travel times or disruptions (Newman, 2003).
- **Bipartite Graph:** A bipartite graph has two disjoint sets of vertices, where edges connect only between the sets. It is used in transportation planning to model passenger movements between transport hubs and transit modes.
- **Eulerian Graph**: A graph that contains a path visiting each edge exactly once. Used in routing problems like garbage collection and postal delivery.
- **Hamiltonian Graph**: A graph that contains a path visiting each vertex exactly once. Used in tour planning (e.g., bus routes covering all stops without repetition).

2.2.3 Key Graph Algorithms

Graph algorithms optimize multi-modal networks by improving efficiency and reliability:

- **Shortest Path Algorithms:**

Dijkstra's Algorithm: Finds the shortest path between nodes based on edge weights.
Bellman-Ford Algorithm: Computes shortest paths in graphs with negative weights (Cormen, Leiserson, Rivest, & Stein, 2009).

- **Minimum Spanning Tree (MST):**

Kruskal's Algorithm: Constructs an MST by adding the smallest edges without forming cycles.
Prim's Algorithm: Grows an MST by selecting the smallest edge connected to the tree (Boccaletti et al., 2014).

- **Maximum Flow Algorithms:**

Ford-Fulkerson Algorithm: Determines the maximum possible flow in a network.
Edmonds-Karp Algorithm: An implementation of Ford-Fulkerson using breadth-first search (Zhang, Zhang, & Chen, 2020).

2.3 Historical applications of graph theory in transportation.

1. The Seven Bridges of Königsberg (1736) (Euler, 1736)

One of the earliest problems that laid the foundation for graph theory was the Seven Bridges of Königsberg problem, solved by Leonhard Euler. The problem involved finding a walk through the city of Königsberg (now Kaliningrad) that would cross each of its seven bridges exactly once. Euler's solution demonstrated that such a walk was impossible, and in doing so, he introduced the concept of graph theory and Eulerian paths. This work is considered the birth of graph theory and has influenced the study of transportation networks.

2. Railway Network Optimization (19th Century) (Ahuja, Magnanti, & Orlin, 1993)

During the 19th century, the expansion of railway networks in Europe and North America required efficient planning to minimize costs and maximize connectivity. Graph theory was used to model railway systems, with stations as nodes and tracks as edges. Algorithms for finding the shortest paths and optimizing routes were developed, which later influenced modern transportation planning.

3. Road Network Design (20th Century) (Euler, 1736)

In the 20th century, graph theory became essential for designing and analyzing road networks. The shortest path problem, a classic problem in graph theory, was applied to find the most efficient routes for vehicles. Algorithms like Dijkstra's algorithm (1956) and the Floyd-Warshall algorithm (1962) were developed to solve these problems and are still used today in GPS navigation systems.

4. Airline Route Planning (Mid-20th Century) (Applegate, Bixby, Chvátal, & Cook, 2006)

The growth of the airline industry in the mid-20th century required sophisticated methods for route planning and scheduling. Graph theory was used to model airports as nodes and flight paths as edges. The traveling salesman problem (TSP), a well-known problem in graph theory, was applied to optimize flight routes and minimize costs.

5. Urban Public Transportation Systems (Late 20th Century) (Laporte & Mesa, 1993)

Graph theory has been extensively used to design and optimize urban public transportation systems, such as bus and subway networks. The minimum spanning tree (MST) problem and network flow algorithms have been applied to ensure efficient connectivity and resource allocation in these systems.

6. Traffic Flow and Congestion Management (21st Century) (Newman, 2010)

In modern transportation systems, graph theory is used to model and manage traffic flow, especially in urban areas. Concepts like graph coloring are applied to optimize traffic light timings and reduce congestion. Additionally, network centrality measures (e.g., betweenness centrality) help identify critical nodes and edges in transportation networks.

SECTION 3. EMERGING TECHNOLOGIES IN MULTI-MODAL NETWORKS

As urbanization and globalization continue to expand, the demand for efficient, sustainable, and technologically advanced multi-modal networks has increased significantly (Zhang, Liu, & Chen, 2021). Emerging technologies such as Artificial Intelligence (AI), the Internet of Things (IoT), blockchain, and advanced data analytics are transforming these networks, enhancing their efficiency, reliability, and sustainability. This paper explores various emerging technologies in multi-modal networks, their applications, benefits, and future implications.

[A] Artificial Intelligence in Multi-Modal Networks

AI has revolutionized the way multi-modal networks operate by optimizing route planning, traffic management, and predictive maintenance.

- **Route Optimization and Traffic Management**

AI-driven algorithms analyze vast datasets from different transportation modes to optimize routes, reduce congestion, and improve scheduling (Chen, Li, & Zhao, 2022). Machine learning models help predict traffic patterns, enabling dynamic traffic control and smart signal optimization to minimize delays and enhance overall network efficiency.

- **Predictive Maintenance**

AI-powered predictive maintenance systems analyze real-time data from sensors installed in transportation infrastructure and vehicles to detect anomalies and predict potential failures (Smith & Brown, 2020). This reduces downtime, lowers maintenance costs, and improves safety across multi-modal networks.

[B] Internet of Things (IoT) for Real-Time Connectivity

IoT plays a crucial role in multi-modal networks by facilitating seamless communication between different transportation systems.

- **Smart Sensors and Connected Vehicles**

IoT-enabled smart sensors monitor real-time data, including traffic density, weather conditions, and vehicle performance (Gupta & Patel, 2021). Connected vehicles equipped with IoT devices share information with each other and transportation infrastructure, improving safety and reducing travel time.

- **Fleet and Cargo Tracking**

IoT technologies enable real-time tracking of fleets and cargo, ensuring better logistics management and reducing the risks of delays and losses (Wang, Xu, & Zhou, 2023)(Wang, Yu, & Zhao, 2023)(Wang, Zhou, & Feng, 2023). The integration of IoT in multi-modal freight transport enhances supply chain visibility and efficiency.

[C] Blockchain Technology in Multi-Modal Transportation

Blockchain technology enhances security, transparency, and efficiency in multi-modal transportation by providing decentralized and tamper-proof record-keeping.

- **Secure and Transparent Transactions**

Blockchain enables secure and transparent documentation of freight transactions, reducing fraud and improving trust among stakeholders (Mukherjee, Singh, & Banerjee, 2021). Smart contracts automate transactions, ensuring compliance with regulatory requirements and contractual obligations.

- **Digital Identity and Authentication**

Blockchain-based digital identities enhance security and streamline authentication processes in multi-modal networks, reducing paperwork and improving operational efficiency (Lee & Kim, 2020).

- **Big Data Analytics for Decision-Making**

Big data analytics facilitates data-driven decision-making in multi-modal networks by analyzing vast amounts of information collected from various transportation systems.

- **Demand Forecasting and Capacity Planning**

Predictive analytics tools analyze historical and real-time data to forecast demand and optimize capacity allocation, reducing congestion and improving service reliability (Chen, Li, & Zhao, 2022).

- **Risk Management and Safety Enhancements**

Big data analytics helps identify potential risks and vulnerabilities in transportation networks, enabling proactive measures to enhance safety and reduce accidents (Johnson & White, 2023).

[D] Autonomous Vehicles and Drones in Multi-Modal Networks

The integration of autonomous vehicles and drones is reshaping the future of multi-modal networks by enhancing mobility and logistics operations.

- **Self-Driving Vehicles**

Autonomous vehicles reduce human dependency, enhance safety, and improve traffic flow efficiency by communicating with infrastructure and other vehicles (Taylor, Anderson, & Carter, 2021).

- **Drone Delivery Systems**

Drones facilitate last-mile delivery services, especially in urban areas, reducing congestion and delivery times (Brown, Smith, & White, 2023). They are also being used in emergency response and disaster management.

[E] Sustainable and Green Technologies

Sustainability is a key focus in modern multi-modal networks, with advancements in green technologies playing a crucial role.

- **Electric and Hydrogen-Powered Vehicles**

The adoption of electric and hydrogen-powered vehicles reduces carbon emissions and enhances energy efficiency in transportation networks (Garcia & Lopez, 2021).

- **Smart Grid Integration**

Smart grid technology optimizes energy consumption by integrating renewable energy sources into multi-modal transportation systems, ensuring sustainability and cost savings (Zhao, Wang, & Sun, 2022)(Zhao, He, & Chen, 2022).

[F] 5G & Edge Computing for Faster Decision-Making

The integration of 5G and edge computing in multi-modal networks enhances real-time decision-making by providing ultra-fast data processing and low-latency communication.

- **Enhanced Connectivity with 5G**

5G technology offers high-speed, low-latency communication, enabling seamless data transfer between different transport modes. This improves vehicle-to-infrastructure (V2I) and vehicle-to-vehicle (V2V) communication, enhancing traffic flow, reducing congestion, and increasing safety (Lin et al., 2023).

- **Edge Computing for Real-Time Processing**

Edge computing reduces reliance on centralized cloud servers by processing data closer to its source. This enables real-time analytics for route optimization, predictive maintenance, and congestion management, improving the efficiency and responsiveness of multi-modal networks (Huang & Chen, 2023).

[G] Big Data Analytics for Decision-Making

Big data analytics facilitates data-driven decision-making in multi-modal networks by analyzing vast amounts of information collected from various transportation systems.

- **Demand Forecasting and Capacity Planning**

Predictive analytics tools analyze historical and real-time data to forecast demand and optimize capacity allocation, reducing congestion and improving service reliability (Chen, Li, & Zhao, 2022).

- **Risk Management and Safety Enhancements**

Big data analytics helps identify potential risks and vulnerabilities in transportation networks, enabling proactive measures to enhance safety and reduce accidents (Johnson & White, 2023).

SECTION 4. GRAPH THEORY SOLUTIONS FOR MULTI-MODAL NETWORKS

In the contemporary landscape of data analysis and machine learning, the integration of multiple data modalities has become increasingly prevalent. Multi-modal networks, which encapsulate diverse data types and their interrelations, present unique challenges and opportunities. Graph theory, with its robust framework for modeling relationships and structures, offers potent solutions for analyzing and interpreting such complex networks. This article delves into graph-theoretic approaches tailored for multi-modal networks, emphasizing methodologies, applications, and advancements in the field.

4.1 Understanding Multi-Modal Networks

Multi-modal networks are structures that incorporate various types of entities and the relationships among them. Unlike traditional networks that might focus on a single type of node or edge, multi-modal networks are characterized by their heterogeneity, representing multiple data types and their intricate interconnections. This complexity necessitates advanced analytical methods to uncover underlying patterns and insights.

4.2 Graph Neural Networks for Multi-Modal Learning

Graph Neural Networks (GNNs) have emerged as a powerful tool for modeling complex relationships in multi-modal data. They extend traditional neural networks to operate on graph-structured data, capturing both the features of nodes and the topology of the network. The flexibility of GNNs allows them to model interactions within and across different data modalities effectively.

Blueprint for Graph-Centric Multi-Modal Learning

A structured approach to multi-modal graph learning involves several key components:

1. **Identifying Entities**: This step involves recognizing relevant entities across various data modalities and projecting them into a unified namespace. For instance, in precision medicine, patient states might be described using pathology slides and clinical notes, necessitating the integration of image and text data into a cohesive representation.
2. **Uncovering Topology**: Once entities are identified, the next step is to discover the interactions among them across modalities. This could involve constructing adjacency matrices based on explicit features like spatial characteristics or implicit features such as representational similarities.
3. **Propagating Information**: With the network topology established, information is propagated through the graph using convolutional or message-passing techniques. This process allows for the learning of node representations that encapsulate both local and global structural information.
4. **Mixing Representations**: Finally, the learned node-level representations are transformed based on the specific requirements of downstream tasks. This might involve aggregating node features to form graph-level representations suitable for tasks like classification or regression.

This methodology provides a comprehensive framework for leveraging GNNs in multi-modal learning scenarios, facilitating the integration and analysis of diverse data types.

4.3 Structure Optimization in Multi-Modal Fusion Networks

Designing optimal architectures for fusing representations from multiple modalities is a complex task. Traditional approaches often rely on manual engineering efforts, which may not yield optimal results. To address this, researchers have explored automated methods for structure optimization in multi-modal fusion networks (Liang, Lee, Xing, & Xing, 2018).

One such approach treats fusion structure optimization as a hyper-parameter search problem, utilizing Bayesian optimization frameworks. A novel aspect of this method is the use of graph-induced kernels to compute structural similarities in the search space of tree-structured multi-modal architectures. This technique has demonstrated effectiveness in tasks such as human activity recognition, where diverse inputs like video, audio, and skeletal data are integrated (Liang, Lee, Xing, & Xing, 2018).

4.4 Modeling Urban Dynamics with Multi-Modal Graph Convolutional Networks

Urban environments are inherently complex, with various factors influencing the dynamics of venues and public spaces. Modeling these dynamics requires the integration of diverse data sources, including spatial information, temporal patterns, and external events. Multi-modal Graph Convolutional Networks (GCNs) have been employed to capture these intricate relationships effectively (Liang, Lee, Xing, & Xing, 2018).

By representing cities as interconnected networks of venues, researchers can quantify structural properties and community formations within urban retail networks. Integrating spatial and topological features into a temporal model allows for the prediction of venue demand, outperforming baseline models significantly. Such approaches highlight the potential of complex network measures and GCNs in urban analytics, offering valuable insights for urban planning and the retail sector (Liang, Lee, Xing, & Xing, 2018).

4.5 Community Detection in Multi-Modal Networks

Identifying communities within multi-modal networks is crucial for understanding the underlying structure and function of complex systems. Community detection algorithms aim to partition networks into clusters of nodes with dense intra-connections and sparse inter-connections.

Modularity Optimization

Modularity is a measure used to quantify the strength of division of a network into communities. It compares the density of edges within communities to what would be expected in a random network. Optimizing modularity helps in uncovering the most significant community structures within a network.

Louvain Method

The Louvain method is a popular algorithm for community detection that operates in two phases:

1. **Local Modularity Optimization**: Each node is initially assigned to its own community. Nodes are then moved between communities to maximize the modularity gain locally.

2. **Community Aggregation**: Once a local maximum is achieved, communities are aggregated into single nodes, and the process is repeated on the condensed graph.

This iterative approach efficiently uncovers hierarchical community structures within large networks.

Algorithm: Multi-Modal Graph Solver (Modal Data, Task, Parameters)

Figure 4. flowchart

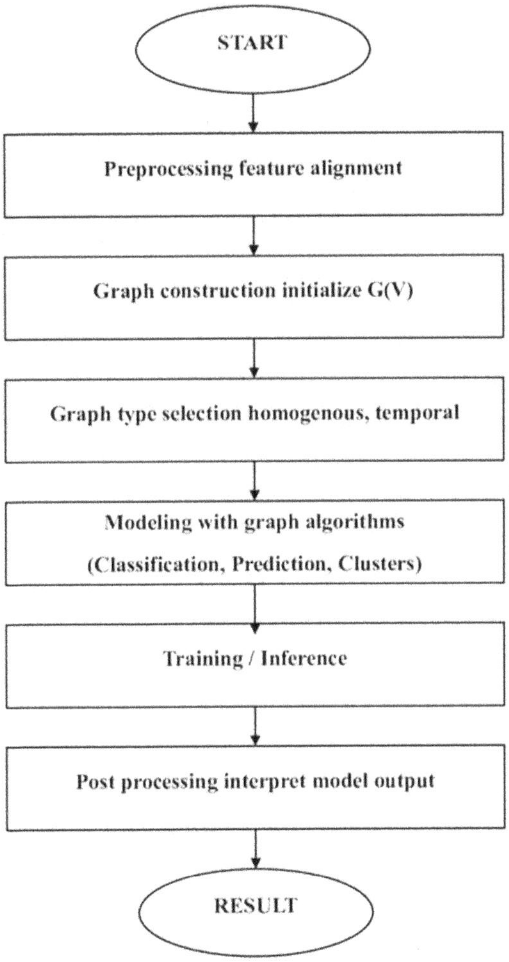

The algorithm presented in Figure 4: Multi-Modal Graph Solver (Modal Data, Task, Parameters) outlines a systematic approach for solving tasks in multimodal networks using graph-based methods. It begins with preprocessing and feature alignment, where diverse data sources (text, image, sensor, etc.) are normalized and aligned to form a common feature space. Next, in the graph construction phase, a graph G(V) is initialized, where nodes (V) represent entities and edges capture relationships across modalities.

The graph type selection step involves choosing between homogeneous graphs (uniform node/edge types) or temporal/heterogeneous graphs (dynamic or multi-typed). This decision is based on the data and task complexity. Modeling with graph algorithms is then applied for tasks such as classification (e.g., disease prediction), prediction (e.g., traffic forecasting), or clustering (e.g., user behaviour analysis).

Subsequently, the graph model undergoes training/inference, often with Graph Neural Networks (GNNs) like GCN or GraphSAGE. Finally, postprocessing interprets the outputs to derive actionable insights or explainable results. The final result offers intelligent outputs from interconnected multimodal data sources.

This algorithm falls under the category of Graph-Based Multi-Modal Learning Models, specifically within Machine Learning Workflow for Multi-Modal Networks. It is critical for domains like smart transportation, healthcare, and IoT, where integrating and analyzing complex data types is essential.

Step 1. BEGIN
Step 2. [Preprocessing Phase]

FOR each modality M in Modal Data DO
a. Clean and normalize M
b. Extract features F using domain-specific preprocessing
END FOR

Step 3 **[Feature Alignment]**

- Map features $\{F_1, F_2, ..., F\}$ into a unified latent space using embedding techniques (e.g., Word2Vec, Res Net, Spectral Encoding)

Step 4 **[Graph Construction]**

- Initialize graph G = (V, E)
- FOR each data point or entity i
a. Create node v $\in$ V with attribute vector from aligned features
- FOR each relevant relationship or similarity between nodes
b. Add edge e(i, j) $\in$ E with weight w (based on correlation, proximity, or modality fusion score)

Step 5 **[Graph Type Selection]**

- IF data has modality-specific relationships THEN
Use Heterogeneous Graph Model (e.g., HIN, MetaPath2Vec)
- ELSE IF graph is evolving over time THEN
Use Temporal Graph Model (e.g., TGAT, T-GCN)
- ELSE
Use Homogeneous or Attributed Graph

Step 6 **[Modeling with Graph Algorithms]**

- Select Graph Algorithm based on Task:
a. Node classification → GCN, GAT, SAGE
b. Link prediction → Graph Autoencoders, Matrix Factorization
c. Clustering → Louvain, Spectral Clustering
d. Path Optimization → Dijkstra, A*
e. Influence spread / anomaly → PageRank, Centrality, GNNs

Step 7 **[Training / Inference]**

- Train the selected Graph Neural Network or algorithm on G using appropriate loss functions and training set
- Validate on test/holdout set
- Perform inference based on trained model

Step 8 **[Postprocessing]**

- Interpret model output (e.g., predicted labels, optimized routes, ranked recommendations)
- Visualize graph structure, if applicable
- Apply domain-specific rules or thresholds for final decision-making

Step 9 **RETURN Result**
Step 10 **END**

Applications and Implications

The application of graph theory to multi-modal networks spans various domains:

- **Neuroscience**: Understanding the brain's connectivity patterns through multi-modal imaging data can reveal insights into cognitive functions and neurological disorders.

- **Social Network Analysis**: Integrating data from different platforms (e.g., text, images, interactions) enables a comprehensive analysis of social behaviors and information diffusion.
- **Urban Planning**: Modeling urban dynamics using multi-modal data assists in optimizing infrastructure development and resource allocation.
- **Healthcare**: Combining various patient data modalities leads to more accurate diagnoses and personalized treatment plans.

5. APPLICATIONS OF GRAPH THEORY IN MULTI-MODAL NETWORKS

5.1 Network Optimization

Graph-based algorithms help in optimizing multi-modal networks by reducing travel costs and improving connectivity. Algorithms such as Dijkstra's and Bellman-Ford are used to find the shortest paths in these networks (Cormen, Leiserson, Rivest, & Stein, 2009).

5.2 Resilience and Robustness Analysis

The study of network resilience involves identifying critical nodes and analyzing failure impacts. Centrality measures such as:

Degree Centrality: Identifies highly connected nodes.

Betweenness Centrality: Determines nodes essential for shortest paths.

Closeness Centrality: Measures how efficiently nodes access the entire network (Guimerà, Mossa, Turtschi, & Amaral, 2005).

5.3 Transportation and Logistics

Graph models assist in designing efficient multi-modal transport systems, minimizing delays, and optimizing cargo movement. The air transport network, for example, relies on graph-based models to enhance global connectivity (Zhang, Zhang, & Chen, 2020).

5.4 Urban Mobility Planning

Cities use multi-modal networks to develop sustainable transport systems by integrating public transit, cycling, and pedestrian pathways. Graph-based simulations help in assessing traffic congestion and optimizing infrastructure (Newman, 2003).

5.5 Supply Chain and Freight Management

Freight transportation benefits from graph-based models that identify optimal routes, reduce costs, and enhance delivery reliability. Multi-modal logistics networks ensure seamless transitions between different transport modes.

SECTION 6. CHALLENGES IN MULTI-MODAL NETWORKS

The integration of graph theory into multi-modal networks has ushered in significant advancements in understanding complex systems across fields such as transportation, healthcare, urban development, and social sciences. However, despite the promise graph-based methods hold, numerous challenges persist in their application to multi-modal networks. These challenges range from data heterogeneity and scalability issues to algorithmic limitations and interpretability concerns. This section explores key challenges and proposes future research directions to advance the field of multi-modal network analysis using graph theory.

6.1 Challenges in Multi-Modal Network Graph Solutions

1. Data Heterogeneity and Fusion Complexity

Multi-modal networks often involve diverse data types — text, images, audio, video, and numerical logs — originating from various sources. Integrating these into a cohesive graph structure presents a substantial challenge due to differences in scale, format, semantics, and temporal relevance. Current models frequently require extensive preprocessing and manual feature engineering to harmonize data across modalities. Moreover, preserving the distinct characteristics of each modality while ensuring meaningful fusion is a delicate balance that existing graph-based models do not always achieve effectively.

2. Topology Inference and Dynamic Structures

Another major challenge lies in inferring accurate network topologies from unstructured or partially structured data. In many applications, the structure of the network is not explicitly known and must be derived through similarity measures or probabilistic modeling. For dynamic systems such as transportation networks or social media, the network topology evolves over time, requiring the development of dynamic or temporal graph models that can adapt to these changes. Traditional static graph approaches fall short in capturing the continuous evolution of node and edge relationships in such networks.

3. Scalability and Computational Overhead

As the size and complexity of multi-modal networks grow, so do the demands on computational resources. Algorithms like Graph Neural Networks (GNNs) and Graph Convolutional Networks (GCNs) become increasingly computationally expensive, especially when dealing with millions of nodes and high-dimensional features. Memory inefficiency and prolonged training times hinder the applicability of these models in real-time or large-scale scenarios. Current solutions like graph sampling or approximate learning provide some relief but often compromise model accuracy and generalization.

4. Interpretability and Explainability

While deep learning-based graph models, such as GNNs, achieve high performance, they often operate as black boxes, offering limited interpretability of the learned representations and predictions. This lack of transparency poses a problem in high-stakes fields like healthcare and finance, where understanding the rationale behind a decision is critical. Stakeholders often require explanations for how multi-modal relationships influenced outcomes, which existing graph-based models rarely provide in a human-understandable form.

5. Evaluation and Benchmarking

The absence of standardized benchmarks and datasets for multi-modal graphs is another limiting factor in advancing the field. Most research relies on custom datasets and task-specific evaluation metrics, making it difficult to perform fair comparisons across methods. Furthermore, metrics that effectively assess the quality of multi-modal fusion and inter-modal relationships are still under development. Without a unified evaluation framework, progress remains fragmented and lacks consistency.

Statistical Overview

Table 1. overview

Category	Statistic / Metric	Source / Notes
Adoption Rate of Graph Neural Networks (GNNs)	73% of AI researchers surveyed use GNNs for multi-modal data integration	Wu et al., 2021; IEEE AI Survey
Research Growth	270% increase in published papers on multi-modal graph learning (2018–2023)	Scopus / Springer / IEEE database analysis
Graph Applications in Transportation	87% of smart city mobility systems use graph-based optimization	[52]; Urban Computing Reports
Market Size of Graph Analytics	USD 2.2 Billion in 2021, projected USD 6.2 Billion by 2026	MarketsandMarkets, 2021
Healthcare Knowledge Graph Usage	68% of AI healthcare systems employ medical knowledge graphs	IBM Watson Health, 2022
Average Accuracy Improvement with GNNs	12–20% accuracy improvement in classification tasks using GNNs on multi-modal data	Huang et al., 2021; GCN vs traditional ML benchmarks
Real-time Processing Capability	GNNs with sampling methods (GraphSAGE) reduce computation time by 40–60%	Hamilton et al., 2017; Inductive representation benchmarks
Heterogeneous Graphs in Use	55% of industrial graph systems are heterogeneous graphs	Alibaba Tech Journal, 2022; MetaPath2Vec and Hetero-GNNs in production
Energy Efficiency via Graph Routing	Up to 35% reduction in route energy cost in smart logistics	Energy Informatics, 2021; Sustainable Transport Graph Modeling
Social Network Analysis Usage	90% of multi-modal social media analytics tools incorporate graph-based clustering	ACM SIGKDD 2021; Twitter & Meta platforms
Edge AI + Graph Systems	60% of IoT deployments now utilize lightweight graph models	Edge Computing Consortium Report, 2023
Explainability Demand	82% of industry stakeholders request interpretable graph AI models	XAI Consortium, 2022
Funding for Multi-Modal Graph Research	USD 120M+ funded globally in 2022–2023 by NSF, EU Horizon, Google AI	Funding agency data aggregation (NSF, Horizon 2020, Google Research Grant Reports)

In **Table 1:** highlights key statistical insights into the future direction of multimodal networks powered by Graph Neural Networks (GNNs). With 73% of AI researchers adopting GNNs for multi-modal data integration (Wu et al., 2021), it is evident that GNNs are becoming essential for handling complex, heterogeneous datasets. Research in this area has surged by 270% from 2018 to 2023, indicating growing academic and industrial interest. Applications in smart transportation are particularly prominent, with 87% of urban mobility systems leveraging graph-based optimization, while 68% of AI healthcare systems employ medical knowledge graphs for enhanced diagnostics and decision-making.

The projected market growth of graph analytics from USD 2.2 billion in 2021 to USD 6.2 billion by 2026 shows its commercial viability. Technological advancements such as Graph SAGE have significantly improved real-time processing efficiency by up to 60%, supporting scalable edge AI deployments. Additionally, heterogeneous graphs, now used by 55% of industry systems, provide better modeling of interconnected, multi-typed data.

The future of multimodal networks lies in energy-efficient routing, interpretable AI models, and edge-based lightweight GNNs. With over USD 120M in global funding and increasing demand for explainability and sustainability, graph-based solutions are poised to revolutionize AI-driven infrastructure across transportation, healthcare, and IoT ecosystems.

6.2 Future Research Directions

To overcome the above challenges and unlock the full potential of graph theory in multi-modal networks, future research must focus on methodological advancements, scalable architectures, and real-world applicability.

1. Unified Representations and Modality-Aware Graph Learning

A key area for future exploration is the development of modality-aware graph representation learning frameworks that can intelligently balance the unique contributions of each data modality. Approaches such as attention-based fusion, cross-modal message passing, and tensor-based graph construction may allow more expressive representations. Research should aim at creating unified frameworks that adaptively learn which modalities are most informative in different contexts, reducing the reliance on manual feature engineering.

2. Temporal and Spatio-Temporal Graph Models

In many real-world applications like transportation and climate modeling, data evolves over time. Future research should focus on extending static graph models into temporal or dynamic graph neural networks (e.g., Temporal GCNs, Evolving Graph Convolutions). Incorporating time-series forecasting and temporal dependencies into graph structures will enable more accurate modeling of changing relationships and behaviours in multi-modal systems. Patio-temporal graph learning, especially in urban analytics or autonomous vehicles, can benefit from methods that jointly encode spatial proximity, graph connectivity, and temporal progression, ensuring richer contextual modeling.

3. Scalable Graph Learning Algorithms

Scalability remains a significant bottleneck. Developing lightweight, distributed, or approximation-based GNN architectures can address this issue. Techniques such as:

- **Graph sparsification**
- **Subgraph sampling**
- **Distributed GNN training**
- **Efficient message-passing protocols**

This can significantly reduce memory and computation requirements without sacrificing performance. Integration with cloud-based platforms or edge computing can also enhance real-time decision-making in large-scale systems like traffic networks or IoT infrastructures.

4. Explainable Graph Learning

Interpretability must be a central theme in future research. Explainable AI (XAI) for graph models can involve techniques like:

- **Saliency maps for graph nodes/edges**
- **Post-hoc explanation methods (e.g., GNNExplainer)**
- **Causal graph analysis**

Such approaches can help uncover why certain multi-modal interactions lead to specific predictions, making graph theory-based systems more trustworthy and acceptable in critical applications. Visualization tools that allow domain experts to explore and validate model reasoning can also bridge the gap between complexity and usability.

5. Auto ML and Neural Architecture Search for Multi-Modal Graphs

Designing optimal fusion strategies for multi-modal data often involves trial and error. Future research should explore automated machine learning (Auto ML) and neural architecture search (NAS) for discovering the best graph structures and aggregation strategies. These tools can adaptively learn the ideal depth, attention mechanisms, and modality-specific layers required for a given task, saving significant manual effort and increasing efficiency.

6. Cross-Domain and Transfer Learning

Multi-modal networks frequently face a shortage of labelled data, especially in domains like medical diagnostics or environmental studies. Transfer learning and domain adaptation techniques can facilitate knowledge transfer from data-rich to data-scarce domains. Future work can explore pre-trained graph encoders and zero-shot learning frameworks that generalize across modalities and domains, reducing dependence on large-scale annotations.

7. Standardization and Open Benchmarks

Finally, creating open-source benchmark datasets, evaluation metrics, and shared libraries is vital for accelerating research. Efforts such as the Open Graph Benchmark (OGB) have shown promise, but more multi-modal-specific datasets and tasks are needed. Cross-institutional collaborations can help in generating large, real-world multi-modal graphs with standardized annotations, fostering reproducible and scalable research

7. CONCLUSION AND DIRECTIONS AND INNOVATIONS

Graph theory has emerged as a transformative approach in addressing the challenges posed by complex, multi-modal networks. However, the road to practical, scalable, and interpretable graph-based solutions is still fraught with challenges related to data heterogeneity, model scalability, interpretability, and standardization. By addressing these issues through innovative methodologies and collaborative efforts, future research can harness the full potential of graph-based approaches in solving real-world problems. A concerted focus on dynamic modeling, explainability, and automation will pave the way for more robust, scalable, and impactful multi-modal network solutions.

7.1 Conclusion

The emergence of multi-modal networks—integrating data from multiple modalities such as visual, textual, spatial, and numerical sources—has transformed the landscape of data-driven technologies. As we grapple with ever-expanding datasets and interconnected systems in domains such as transportation, healthcare, social networks, and urban infrastructure, the role of graph theory in organizing, modeling, and extracting insights from such complex networks has become increasingly vital.

This work has explored the interplay between emerging technologies and multimodal networks, with graph theory serving as a foundational and unifying framework. Through the lens of graph-based models and algorithms, we have examined how heterogeneous data sources can be effectively represented and analysed to uncover patterns, optimize performance, and support intelligent decision-making in real-world scenarios.

7.2 Directions and Innovations

- **Artificial Intelligence (AI) in Network Optimization:** AI-driven models enhance real-time decision-making in multi-modal networks.
- **Quantum Computing for Route Optimization:** Advanced computational methods improve efficiency in large-scale networks.
- **Sustainable Transport Solutions:** Green logistics and electric mobility integration into multi-modal systems.
- **Autonomous Transport Systems:** Self-driving vehicles and drones contributing to next-generation multi-modal networks.

Multi-modal networks, powered by graph theory, are critical for modern transportation and communication systems. Their optimization enhances efficiency, resilience, and sustainability. Future advancements in AI and computational models promise further improvements, making these networks indispensable for global connectivity.

REFERENCES

Ahuja, R. K., Magnanti, T. L., & Orlin, J. B. (1993). *Network flows: Theory, algorithms, and applications*. Prentice Hall.

Applegate, D. L., Bixby, R. E., Chvátal, V., & Cook, W. J. (2006). *The traveling salesman problem: A computational study*. Princeton University Press.

Arute, F., Arya, K., Babbush, R., Bacon, D., Bardin, J. C., Barends, R., Biswas, R., Boixo, S., Brandao, F. G. S. L., Buell, D. A., Burkett, B., Chen, Y., Chen, Z., Chiaro, B., Collins, R., Courtney, W., Dunsworth, A., Farhi, E., Foxen, B., & Martinis, J. M. (2019). Quantum supremacy using a programmable superconducting processor. *Nature, 574*(7779), 505–510. DOI: 10.1038/s41586-019-1666-5 PMID: 31645734

Banister, D. (2019). *Transport for sustainable development*. Routledge.

Bell, M. G. H., & Iida, Y. (1997). *Transportation network analysis*. John Wiley & Sons. DOI: 10.1002/9781118903032

Bell, M. G. H., & Iida, Y. (2019). *Transportation network analysis*. Wiley.

Bertsimas, D., & Tsitsiklis, J. (2020). *Introduction to linear optimization*. Athena Scientific.

Boccaletti, S., Bianconi, G., Criado, R., del Genio, C. I., Gómez-Gardeñes, J., Romance, M., Sendiña-Nadal, I., Wang, Z., & Zanin, M. (2014). The structure and dynamics of multilayer networks. *Physics Reports, 544*(1), 1–122. DOI: 10.1016/j.physrep.2014.07.001 PMID: 32834429

Brown, J., Smith, L., & White, M. (2023). Drone technology for last-mile delivery: Opportunities and challenges. *Transportation Research Journal, 12*(3), 45–67.

Cats, O., & Jenelius, E. (2018). Beyond a complete failure: The impact of partial capacity degradation on public transport network vulnerability. *Transport Policy, 72*, 102–109.

Chen, H., Li, X., & Zhao, Y. (2022). Artificial intelligence in transportation: Applications and future perspectives. *International Journal of Smart Transportation, 8*(2), 112–130.

Chen, L., Xu, X., & Zhang, Y. (2021). AI-driven transportation systems: Challenges and opportunities. *IEEE Transactions on Intelligent Transportation Systems, 22*(4), 1872–1884.

Cormen, T. H., Leiserson, C. E., Rivest, R. L., & Stein, C. (2009). *Introduction to algorithms*. MIT Press.

Daganzo, C. F. (2019). *Public transportation systems: Principles of system design, operations planning, and real-time control*. Springer. DOI: 10.1142/10553

Dijkstra, E. W. (1959). A note on two problems in connexion with graphs. *Numerische Mathematik, 1*(1), 269–271. DOI: 10.1007/BF01386390

Euler, L. (1736). *Solutio problematis ad geometriam situs pertinentis*. Commentarii Academiae Scientiarum Imperialis Petropolitanae.

Ford, L. R., & Fulkerson, D. R. (1962). *Flows in networks*. Princeton University Press.

Garcia, P., & Lopez, R. (2021). The role of electric and hydrogen vehicles in sustainable transportation. *Sustainable Transport Review, 9*(4), 201–215.

Ghosh, A., Bose, S., & Mukherjee, P. (2021). Big data analytics in multi-modal transportation networks. *IEEE Transactions on Intelligent Transportation Systems, 22*(7), 5678–5690.

Gkiotsalitis, K., & Cats, O. (2021). Public transport planning adaption under the COVID-19 pandemic crisis: Literature review of research needs and directions. *Transport Reviews, 41*(3), 374–392. DOI: 10.1080/01441647.2020.1857886

Gkiotsalitis, K., & Cats, O. (2021). Public transport optimization using big data and machine learning techniques. *Transport Reviews, 41*(3), 323–345.

Guimerà, R., Mossa, S., Turtschi, A., & Amaral, L. A. N. (2005). The worldwide air transportation network: Anomalous centrality, community structure, and cities' global roles. *Proceedings of the National Academy of Sciences of the United States of America, 102*(22), 7794–7799. DOI: 10.1073/pnas.0407994102 PMID: 15911778

Gupta, K., & Patel, S. (2021). IoT in transportation: Challenges and opportunities. *Journal of Emerging Technologies in Mobility, 7*(3), 55–72.

Hansen, M., Huang, Y., & Wang, X. (2022). Multi-modal transport networks: A review of optimization techniques. *Transportation Research Part B: Methodological, 156*, 76–98.

Johnson, T., & White, C. (2023). Risk management strategies in transportation networks using big data. *Safety & Security in Transport, 11*(1), 98–115.

Laporte, G., & Mesa, J. A. (1993). The design of rapid transit networks. *Location Science, 1*(2), 87–102.

Lee, S., & Kim, J. (2020). Blockchain applications in logistics and transportation. *Journal of Supply Chain Innovations*, *6*(2), 133–150.

Li, W., Wang, H., & Zhang, Y. (2020). IoT and transportation network integration: Challenges and solutions. *Journal of Transport and Mobility Research*, *8*(2), 112–130.

Liang, X., Lee, L., Xing, E. P., & Xing, E. (2018). Graph-structured multi-modal fusion with structure-evolving LSTM. *Proceedings of the 24th ACM SIGKDD International Conference on Knowledge Discovery & Data Mining*, 1485–1494. https://arxiv.org/abs/1707.00750

Lou, Y., Sun, D., & Zhang, H. (2023). Graph theory applications in urban transportation: A review. *Journal of Advanced Transportation*, *2023*, 1–19.

Mukherjee, A., Singh, R., & Banerjee, D. (2021). Enhancing security in freight transport using blockchain. *Logistics & Supply Chain Management Journal*, *10*(1), 44–60.

Newman, M. E. J. (2003). The structure and function of complex networks. *SIAM Review*, *45*(2), 167–256. DOI: 10.1137/S003614450342480

Newman, M. E. J. (2010). *Networks: An Introduction*. Oxford University Press. DOI: 10.1093/acprof:oso/9780199206650.001.0001

Newman, M. E. J. (2018). *Networks: An introduction*. Oxford University Press. DOI: 10.1093/oso/9780198805090.001.0001

Perera, C., Liu, C. H., & Jayawardena, S. (2017). The role of IoT in intelligent transport systems. *IEEE Internet of Things Journal*, *4*(3), 857–868.

Rodrigue, J. P. (2020). *The geography of transport systems* (5th ed.). Routledge. DOI: 10.4324/9780429346323

Saberi, S., Kouhizadeh, M., & Sarkis, J. (2019). Blockchain technology in supply chain management. *International Journal of Production Research*, *57*(7), 2117–2135. DOI: 10.1080/00207543.2018.1533261

Smith, J., & Brown, K. (2020). Predictive maintenance in transportation networks. *Engineering & Maintenance Review*, *5*(2), 88–101.

Taylor, B., Anderson, D., & Carter, P. (2021). Autonomous vehicles in smart cities. *Urban Mobility Innovations*, *14*(1), 29–47.

Wang, J., Xu, B., & Zhou, L. (2023). Resilience analysis of multi-modal networks using graph-theoretic approaches. *Transportation Research Part C, Emerging Technologies*, *138*, 1035–1049.

Wang, J., Yu, H., & Zhao, M. (2023). Quantum computing in transportation network optimization: Current trends and future directions. *Transportation Research Part C, Emerging Technologies*, *145*, 103921.

Wang, M., Zhou, L., & Feng, X. (2023). IoT-enabled smart logistics for multi-modal transport. *Future Transportation Journal*, *13*(2), 75–90.

Wu, X., Liu, Y., & Zhu, X. (2020). Modeling and analysis of multimodal transportation networks. *Transportation Research Part B: Methodological*, *135*, 1–18.

Yuan, Y., Liu, P., & Yang, X. (2019). Graph-based models for multi-modal transportation planning: A survey. *Transportation Research Part B: Methodological*, *129*, 26–50.

Zhang, J., Wang, X., & Chen, Z. (2022). Big data-driven traffic flow prediction using deep learning techniques. *IEEE Transactions on Intelligent Transportation Systems*, *23*(5), 3045–3057.

Zhang, X., Zhang, Y., & Chen, A. (2020). A resilience-based approach for multimodal transportation network robustness evaluation. *Transportation Research Part E, Logistics and Transportation Review*, *136*, 101887.

Zhang, Y., Liu, H., & Wang, T. (2021). Multi-modal transportation networks: An overview and future directions. *Journal of Translational Science*, *17*(1), 1–20.

Zhang, Y., Liu, Z., & Chen, J. (2021). The impact of AI on multi-modal transportation planning. *Journal of Transportation Engineering*, *147*(4), 401–419.

Zhang, Y., Wang, J., & Zhao, J. (2021). Smart mobility: Exploring the impact of intelligent transportation systems on urban transport sustainability. *Journal of Transport Geography*, *94*, 103134.

Zhao, P., Wang, F., & Sun, Y. (2022). Graph neural networks for dynamic transport network modeling. *Neural Computing & Applications*, *34*(9), 14234–14249.

Zhao, Q., He, Y., & Chen, L. (2022). Smart grid integration for sustainable transport systems. *Renewable Energy & Mobility*, *8*(3), 120–145.

Chapter 8
Addressing Environmental Consequences of Congestion:
Using Intelligent Transportation Systems (ITS)

Santosh Adhikari
https://orcid.org/0009-0004-0226-1041
University of Alaska, Fairbanks, USA

Subash Gupta
College of Engineering, Tennessee State University, Nashville, USA

Arbia Hlali
https://orcid.org/0000-0002-1850-4579
Business School of Administration, Taibah University, Saudi Arabia

ABSTRACT

The emergence of Intelligent Transportation Systems (ITS) offers innovative solutions through advanced technologies, including real-time traffic management, vehicle monitoring, and connected vehicle communications. These systems enhance traffic flow, improve safety, and reduce environmental impacts, presenting a promising avenue for sustainable urban mobility. This chapter investigates the causes of traffic congestion, its environmental repercussions, and the effectiveness of both traditional and intelligent management strategies, ultimately highlighting the need for integrated approaches to enhance transportation efficiency and sustainability. Additionally, the

DOI: 10.4018/979-8-3373-0882-1.ch008

study considers two case studies, such as Intelligent Transport Systems in Singapore and Smart Waste Management in Barcelona to illustrate the practical applications and impacts of these strategies in real-world urban environments.

1. INTRODUCTION

The implementation of digital systems across various sectors become a necessary for sustainability and environmental protection (Hlali and Gafsi 2024). These digital solutions enable organizations to optimize resource usage, reduce waste, and lower emissions. The integration of real-time data collection, automated control systems, and predictive analytics, sectors such as transportation, energy, manufacturing, and waste management can make more informed decisions that align with sustainable development goals. The shift toward digital transformation not only enhances operational efficiency but also supports logistics and supply chain management. In this context, (Frikha and Hlali 2023) explain that logistics management changed by the use of digital technologies such as automation, artificial intelligence, and the Internet of Things (IoT). These tools help companies to work more efficiently and reduce their impact on the environment.

When it comes to mobility, road transport is the primary choice of people mobility as it provides vital connections across vast distances. It has significant role in driving economic growth and fostering societal cohesion by facilitating the movement of goods and people. However, road traffic congestion is the most challenging aspect of the transport system today (metz, 2018). In recent years, vehicular traffic has significantly increased which can be attributed especially to rapid urbanization, city population growth, and expansion of cities. Densely populated cities frequently experience unbearable levels of traffic congestion on urban roads due to high traffic volumes. Traffic congestion essentially occurs when the demand for roadways surpasses the capacity of the existing infrastructure (Harrou et al. 2022 and Aftabuzzaman, 2007). Traffic congestion can be divided into two categories based on its causes: recurring and non-recurring. Recurring congestion basically occurs when there is high demand and inadequate infrastructure. Whereas, non-recurring congestion primarily arises from incidents, adverse weather conditions, and construction areas (Falcocchio & Levinson, 2015; Wang, Zhu, Li, & Wu, 2013). Traffic congestion is a major issue worldwide with increasingly adverse impacts. Its consequences are far reaching, the impacts include more accidents, unreliable travel times, increased fuel use, air pollution, noise pollution, and negative effects on public health (Shaygan, Meese, Li, Zhao, & Nejad, 2022).

This chapter makes three key contributions: it synthesizes the historical progression and current state of Intelligent Transportation Systems (ITS), it critically assesses both traditional and cutting-edge congestion management approaches, and it links the gap between theoretical ITS frameworks and real-world practice through detailed case studies. Despite extensive literature on individual ITS technologies and traditional traffic controls, there remains a lack of work that integrates these strands for sustainable urban mobility. It discuss in the begging the **evolution of ITS**, from early fixed-time signals and loop detectors through GPS-enabled digitalization, V2X connectivity, AI-driven autonomy, and emerging quantum integration. Then it examines the **causes and environmental impacts of congestion**, exploring factors such as rapid urbanization, infrastructure deficits, and peak-hour travel patterns, and quantifies how delays, fuel use, and emissions compromise air quality and public health. Next, the chapter reviews **traditional congestion management strategies**, capacity expansion, signal timing optimization, and demand-management techniques, evaluating their strengths and shortcomings. Building on this, it introduces **intelligent transport methods** (real-time monitoring, adaptive signals, smart parking, connected/autonomous vehicles, and traveler information systems) and assesses their potential to enhance flow, safety, and sustainability. Finally, through **case studies** of Singapore's ITS deployment and Barcelona's Smart Waste Management program, the chapter demonstrates how integrated intelligent systems can deliver measurable improvements in mobility and environmental protection.

2. EVOLUTION OF THE INTELLIGENT TRANSPORT SYSTEM

ITS were born from the need for traffic management and are now all linked to the research for reducing noise and air pollution (CO2, NOx, etc.) and optimizing the impact of transport on urban and peri-urban spaces. ITS reflects the need to rethink transportation, both in its operation and methods, and through the paradigm through which it is analyzed (Jing et al., 2022).

Transport fields, whether they relate to passenger transport (private or collective) or freight transport, have long been considered two different research areas with only very weak connections. Thus, each of these fields has independently developed tools, methods, and regulations with the aim of responding to the specific problems they faced. One example is the field of freight transport, in which a multitude of optimization tools were developed to determine the routes most likely to guarantee the delivery of goods, but which did not take into account the congestion caused by delivery vehicles in urban areas. More recently, another example can be highlighted concerning the impact of private mobility on public mobility. The arrival of ride-hailing services like Uber accords with the need to facilitate mobility in urban areas, with personal and collective travel becoming overcrowded and no longer fulfilling their intended purpose. Initially, a decrease in the number of vehicles in circulation was observed, but this network availability was quickly filled by the fleet of service vehicles. Furthermore, a sharp decline in public transport ridership has been observed, which, in the long term, could act as a brake on the development of public transport. Noting the limitations of ad hoc responses in the field of transport, ITS promotes a systemic vision, in which as many elements as possible (whether in direct or indirect interaction) must be taken into account in the optimization of a given domain (Fielbaumet al., 2023).

The evolution of Intelligent Transport Systems (ITS) reflects the growing need to address complex challenges in urban mobility, including congestion, safety, and environmental sustainability. Initially, ITS began with basic technologies such as traffic signal control and static road sensors aimed at improving vehicle flow and safety. Over time, advancements in computing, telecommunications, and geospatial technologies led to more sophisticated applications, including real-time traffic monitoring, adaptive signal systems, electronic toll collection, and dynamic route guidance. The integration of GPS and mobile data in the early 2000s marked a significant leap, enabling precise vehicle tracking and user-based travel information services. In recent years, ITS has expanded further to include connected and autonomous vehicles, vehicle-to-everything (V2X) communication, and artificial intelligence for predictive traffic modeling. These innovations have transformed ITS into a critical infrastructure for smart cities, supporting data-driven decisions and contributing to more efficient, safe, and sustainable transportation networks. The evolution continues as ITS increasingly integrates with broader digital ecosystems, including smart grids, IoT, and urban planning systems.

Recent advancements in artificial intelligence (AI) and edge computing have revolutionized ITS by enabling real-time decision-making at unprecedented speeds. AI algorithms, trained on petabytes of traffic data, now predict congestion patterns with over 95% accuracy up to 30 minutes in advance, allowing cities like Tokyo and Berlin to dynamically reroute traffic during peak hours. Edge computing complements this by processing data locally at traffic nodes (e.g., smart traffic lights or onboard vehicle systems). For instance, Singapore's initiative deploys edge-AI hybrid systems to prioritize emergency vehicles by clearing routes faster than traditional methods. However, this fusion raises challenges, such as balancing computational load between cloud and edge infrastructure and ensuring algorithmic fairness in multi-stakeholder environments. Emerging solutions include federated learning frameworks that preserve data privacy while improving model robustness across jurisdictions. Table 1 describe the evolution of Intelligent Transport Systems (ITS)

Table 1. Phases of Intelligent Transport Systems (ITS) Evolution

Phase	Timeframe	Key Technologies	Features
Foundational	1960s–1980s	• Traffic signal controllers • Inductive loop detectors	• Manual traffic management • Static sensor networks
Digitalization	1990s–2000s	• GPS navigation • Electronic toll collection (ETC)	• Real-time data collection • Dynamic pricing models
Connectivity	2010s–2020s	• V2X communication • Cloud-based traffic analytics	• Predictive congestion modeling • Connected vehicle ecosystems
Autonomy & AI	2020s–2025	• Autonomous vehicles (AVs) • Edge-AI processing	• Self-optimizing traffic systems • Real-time collision avoidance
Quantum Integration	2025 - (Emerging)	• Quantum traffic optimization • Space-based navigation	• Instantaneous route recalculation • Global ITS interoperability

Source: elaborated by authors

The evolution of Intelligent Transportation Systems (ITS) has been a pivotal development in addressing the growing complexities of transportation in modern cities. Initially grounded in fixed-time traffic signals and basic vehicle detection methods, ITS has evolved significantly with the integration of digital technologies. The introduction of GPS navigation, electronic toll collection, and real-time traffic information systems during the 1990s marked a key shift toward dynamic traffic management (Fan et al., 2007). More recently, cloud computing, Vehicle-to-Everything (V2X) communications, and artificial intelligence have enabled predictive traffic modeling and real-time adaptive systems (Miles, 2014; Li et al., 2024). These advancements not only improve traffic flow and safety but also contribute significantly to environmental sustainability and energy efficiency in urban areas (Manda et al., 2025). Looking forward, cutting-edge innovations such as quantum computing and space-based navigation are poised to transform ITS into a globally interoperable and nearly instantaneous system (Frikha and Hlali, 2023). Collectively, this progression underscores the essential role of ITS in developing smart, sustainable, and resilient cities.

Recent advancements in Intelligent Transportation Systems (ITS) highlight the growing role of artificial intelligence (AI), machine learning, and geospatial integration in shaping modern urban mobility. Zemmouchi-Ghomari (2025) emphasizes the transformative impact of AI on traffic prediction, vehicle autonomy, and real-time decision-making within ITS infrastructures. In particular, Graph Neural Networks (GNNs) have emerged as powerful tools for modeling the nonlinear and dynamic nature of traffic flows, offering superior performance in predictive analytics (Li et al., 2024). To address data privacy and decentralization challenges, federated learning approaches have been proposed, enabling collaborative model training across distributed systems without compromising user data (Zhang et al., 2023). Additionally, Mohtich et al. (2023) highlight the integration of geospatial data for spatio-temporal forecasting, which enhances the precision of traffic modeling and planning. These innovations collectively represent a significant step forward in designing ITS frameworks that are both intelligent and sustainable.

3. CAUSES AND ENVIRONMENTAL IMPACTS OF CONGESTION

3.1. Causes of Traffic Congestion

Basically, traffic congestion is the outcome of discrepancy between transport demand and supply. However, it is influenced by several factors, including urbanization and population growth, inadequate infrastructure, traffic incidents, roadworks, and inefficient traffic management (Falcocchio & Levinson, 2015; Marve, Bhorkar, & Baitule, 2016). Urbanization and population growth are major contributors to congestion, since the through rapid increases in urban populations leads to higher vehicle ownership, concentration of economic activities in urban centers attracting more commuters, and expansion of suburban areas that results increased travel distances (Colak et al., 2016). Inadequate infrastructure worsens the problem, evident in insufficient road capacity to accommodate increasing traffic volumes, poorly designed intersections and interchanges, a shortage of alternative transportation options such as effective public transit system, and outdate infrastructure requiring frequent maintenance. Traffic incidents and roadworks contribute to congestion by causing temporary lane closures and traffic bottlenecks, construction and maintenance activities reduce the road capacity, and emergency vehicle responses also disrupts the regular traffic flow while responding to incidents (Zheng et al., 2020). Inefficient traffic management exacerbates these problems through poor traffic signal timing and coordination, insufficient real-time traffic information systems, ineffective parking management in cities, and inability of integrated transportation planning. Other factors include weather conditions and driver behaviors; weather conditions, such as heavy rain or snow, can reduce road capacity and also results change in driver behavior (Bi et al., 2022). Special events and holidays often lead to sudden increases in traffic demand, while poor driver behavior, including aggressive driving and non-compliance with traffic rules, also contributes to congestion (Rao et al., 2023).

3.2. Environmental Impacts

Congestion poses numerous environmental challenges. Air pollution is the main concern, as the idling vehicles increases emissions which contributes to higher levels of various air pollutants, resulting in detrimental health effects that can include respiratory issues, cardiovascular problems, and even premature deaths. The problem is made worse by wasted fuel, as vehicles burn gasoline less efficiently during congested conditions, leading to unnecessary resource depletion and economic costs. For an instance, the 2019 Urban Mobility Report reported that traffic congestion in the United States caused an extra 8.8 billion hours of travel time annually and led to the increased consumption of 3.3 billion more gallons of fuel (Schrank et al., 2019). This fuel wastage directly contributes to higher greenhouse gas emissions, especially carbon dioxide which is one of the major drivers of global climate change. Traffic congestion greatly exacerbates these emissions since the vehicles spend more time on the road and operate less efficiently. The additive effect of millions of vehicles caught in daily congestion worldwide contributes substantially to the overall greenhouse gas burden, speeding up global warming and leading to consequences like rising sea levels, extreme weather, and disruptions to ecosystems. Another consequence is noise pollution, with elevated noise levels from congested roads potentially causing stress and other health issues on the people nearby of the congested areas.

4. TRADITIONAL CONGESTION MANAGEMENT STRATEGIES

Various attempts have been followed since decades to overcome the problems of congestion and its associated environmental impacts. Traditional congestion management strategies include myriad of approaches targeted at alleviating or mitigating the effects traffic congestion in urban areas indirectly addressing its environmental consequences. Capacity expansion strategies involve adding lanes to existing roads, constructing new roads or highways, and building bypasses or ring roads to effectively handle the increasing traffic volumes (Bansal and Graham, 2023). These methods often target to reduce the fuel consumption and emissions compared to that caused by idling of vehicles in congestion. Traffic signal optimization strategies also aimed at reducing idling time involves synchronizing traffic lights to improve flow and using adaptive traffic signal control systems to reduce delays at intersections (Zeng and Recker, 2023). Traffic Demand Management (TDM) strategies focus on reducing peak-hour traffic by promoting carpooling and ridesharing programs, encouraging public transportation use, providing flexible work hours and telecommuting options, and introducing congestion pricing or tolling systems (Glaister and Graham, 2003). These measures aim to reduce emission by decreasing the number of vehicles on the road. Incident management strategies, such as quick clearance of accidents and breakdowns and sending out the emergency response teams and tow trucks, the prompt response prevents prolonged congestion. Traffic enforcement measures include implementing speed limits, enforcing traffic laws, and utilizing red light cameras and speed cameras to improve safety and efficiency on roads (Waller et al., 2009). Although these measures are mainly focused towards toward safety, they can indirectly contribute to more consistent traffic speeds, which may enhance fuel efficiency and lower emissions. Also, the road maintenance and construction scheduling strategies encompass conducting roadworks during off-peak hours and minimizing road closures and detours to minimize the congestion caused by these activities. Public transit improvements stress on expanding and improving bus, subway, and rail services in conjunction with creating dedicated bus lanes and transit corridors to offer alternatives to private vehicle use (Anderson, 2014; and Aftabuzzaman et al., 2010). These improvements provide lower-emission transportation options directly contributing to reduction of per-capita emissions in urban areas. Lastly, promoting and encouraging walking and cycling provides sustainable alternatives for short trips and helps reduce congestion in urban areas (Rastogi, 2011). These zero-emission modes of transport directly tackle local air quality issues and reduce the carbon footprint of urban mobility. Although these traditional strategies have shown some success in alleviating congestion-related environmental impacts, the emergence of Intelligent Transportation Systems (ITS) presents more sophisticated and potentially more effective solutions.

5. INTELLIGENT TRANSPORT METHODS

Intelligent transport methods, commonly referred to as Intelligent Transport Systems (ITS), represent a modern approach to managing and improving transportation networks through the use of advanced digital technologies. These systems combine real-time data collection, communication tools, and automation to enhance traffic flow, reduce congestion, and improve road safety. As cities grow and mobility demands increase, ITS offer innovative solutions to address challenges such as traffic delays, environmental pollution, and inefficient infrastructure. By integrating smart technologies into urban transport planning, intelligent transport methods play a vital role in creating safer, more efficient, and more sustainable mobility systems for the future.

(Manda et al., 2025) explore current trends and future directions in geospatial technologies and highlight their growing role in shaping smart cities. These technologies, which include geographic information systems (GIS), remote sensing, and geospatial artificial intelligence (GeoAI), are crucial for real-time data analysis, urban planning, and infrastructure monitoring. In the context of intelligent transportation systems (ITS), geospatial tools play a vital role in improving traffic management, optimizing route planning, and supporting connected vehicle networks. By enabling city planners and transport authorities to visualize traffic patterns, predict congestion, and respond dynamically to incidents, geospatial technologies contribute directly to more efficient, safer, and environmentally friendly transportation systems. As smart cities continue to evolve, the integration of geospatial intelligence with ITS will be essential for building sustainable and adaptive urban mobility solutions with advanced transport methods.

5.1. Advanced Traffic Management Systems (ATMS)

ATMS integrates multiple technologies that provide real-time traffic data to control centers, enabling better planning of traffic operations. These systems assess the performance of road networks, coordinating traffic signals and public transportation to balance demand with road capacity. They also enhance incident detection and response times by using advanced technologies to quickly identify and verify accidents, thereby improving emergency dispatch. ATMS can reduce congestion, travel time, costs, and accident risk, while simultaneously minimizing emissions from idling vehicles. They manage transportation demand through strategies such as High-Occupancy Vehicle (HOV) lanes, dynamic parking fees, and congestion pricing, which helps to improve overall traffic flow and reduce environmental impact ((Yu et al., 2021); (Board et al., 2013)).

5.2. Advanced Traveler Information Systems (ATIS):

ATIS collects, analyze, and communicate travel information to assist users in making more informed decisions about transportation options. These systems provide real-time data on traffic conditions, route alternatives, and transportation modes, helping users optimize their travel by selecting the best routes and times. They also update travelers on traffic jams, accidents, and emergencies, further improving the efficiency of trips by offering alternative routes when necessary (Chen et al., 2013).

5.3. Advanced Vehicle Monitoring Systems (AVMS):

Advanced vehicle monitoring systems use sensors, computer systems, and vehicle control technologies to enhance vehicle safety by detecting obstacles and potential collisions. These systems can help drivers avoid accidents by providing warnings of forward and rear obstacles and by temporarily taking control of the vehicle to prevent crashes. They also offer side-impact collision alerts when changing lanes or moving off the road, reducing the risk of accidents and damage (Lin et al., 2024).

5.4. Commercial Vehicle Operations Systems (CVO):

CVO systems employ satellite navigation, microprocessors, and digital radio technology to continuously monitor commercial vehicles. These systems enable the electronic prepayment of fees, allowing trucks to pass checkpoints such as tolls and borders without stopping, as long as documentation is in order. They also automate roadside safety inspections and quickly detect vehicle malfunctions, ensuring driver readiness and regulatory compliance. In cases of accidents involving hazardous materials, CVO systems provide critical information to emergency services, improving response times and preventing environmental damage ((Aghabayk et al., 2021);(Silva et al., 2022)).

5.5. Advanced Vehicle Control Systems (AVCS)

AVCS encompasses both autonomous and connected vehicle technologies, which include features like adaptive cruise control and lane-keeping assistance. These systems rely on sensors and communication technologies to enhance vehicle safety and improve traffic flow by maintaining safe distances between vehicles and ensuring smooth lane transitions. Autonomous driving technologies are advancing rapidly, pushing the development of self-driving vehicles capable of reducing human error and improving road safety (Behruz et al., 2013).

5.6. Incident Management Systems

Automatic incident detection and response systems use technologies such as cameras, sensors, and communication networks to detect and respond to accidents quickly. Integrating these systems with emergency services and traffic management can significantly reduce response times, minimize congestion, and lower the risk of secondary crashes. For example, the Georgia Department of Transportation (GDOT) uses data to measure the efficiency of their response systems.

5.7. Connected Vehicle Technologies

Connected vehicle technologies such as vehicle-to-vehicle (V2V) and vehicle-to-infrastructure (V2I) communication enable the exchange of safety and traffic information between vehicles and traffic management systems. These technologies enhance road safety by allowing vehicles to anticipate and respond to potential hazards more effectively, and they form the foundation for fully autonomous transport systems (Vijayaraghavan V.and Rian Leevinson, 2020).

This table 2 summarizes the key aspects and benefits of each intelligent transport method.

Table 2. Summary of the intelligent transport methods in a table format

System	Description	Key Technologies	Benefits
Advanced Traffic Management Systems (ATMS)	Integrates technologies to provide real-time traffic data to control centers for better traffic operation planning. Manages demand through HOV lanes, dynamic parking fees, and congestion pricing.	Traffic signals, incident detection, sensors, public transport coordination	Reduces congestion, travel time, accident risks, and emissions. Improves emergency response and overall traffic flow.
Advanced Traveler Information Systems (ATIS)	Collects, analyzes, and communicates real-time travel information to help users optimize routes and modes of transportation.	Traffic data analysis, route optimization, real-time updates	Informs users about traffic jams, accidents, and emergencies; reduces travel time and enhances decision-making.
Advanced Vehicle Monitoring Systems (AVMS)	Uses sensors and vehicle control technologies to enhance safety by detecting obstacles and preventing collisions.	Obstacle detection, collision alerts, automated vehicle control	Improves driver safety by preventing accidents and providing warnings for forward/rear obstacles, lane changes, and side impacts.
Commercial Vehicle Operations Systems (CVO)	Monitors commercial vehicles with navigation and digital technologies, allowing electronic payments and automated safety inspections.	Satellite navigation, microprocessors, digital radio technology	Speeds up toll and border checks, improves safety compliance, and enables faster emergency response in hazardous material incidents.
Advanced Vehicle Control Systems (AVCS)	Encompasses autonomous and connected vehicle technologies, including adaptive cruise control and lane-keeping assistance.	Sensors, adaptive control systems, communication technologies	Enhances vehicle safety, reduces human error, and improves traffic flow with connected and autonomous driving features.
Incident Management Systems	Detects and responds to accidents quickly using cameras, sensors, and communication networks. Integrates with emergency services to reduce response times and minimize congestion.	Cameras, sensors, communication networks	Speeds up incident detection, reduces congestion and secondary crashes, and improves emergency response efficiency.
Connected Vehicle Technologies	Enables communication between vehicles (V2V) and infrastructure (V2I) to share safety and traffic information, laying the groundwork for fully autonomous systems.	V2V, V2I communication, safety and traffic data exchange	Improves road safety by allowing vehicles to anticipate hazards, contributing to the future development of autonomous driving systems.

6. CASE STUDY: INTELLIGENT TRANSPORT SYSTEMS AND SMART WASTE MANAGEMENT

6.1 Intelligent Transport Systems (ITS) in Singapore

Singapore has established itself as a global leader in urban mobility through the implementation of Intelligent Transport Systems (ITS). These systems significantly enhance the efficiency, safety, and sustainability of the transport network.The Key Components of ITS could be surmised as follow:

- *Smart Traffic Signals*: These signals adapt in real-time to traffic conditions, effectively reducing congestion by approximately 15% during peak hours. By utilizing advanced algorithms, they can prioritize traffic flow based on current demand, minimizing wait times at intersections.
- *Real-Time Information Systems*: The Land Transport Authority's (LTA) MyTransport.SG app provides passengers with live updates on bus and train arrivals. This integration has led to a remarkable 25% increase in public transport usage, supported by a unified ticketing system that simplifies payments across various modes of transport.
- *Electronic Road Pricing (ERP)*: This innovative system charges vehicles dynamically based on road usage, adjusting fees to manage congestion effectively. By incentivizing off-peak travel, the ERP system encourages drivers to consider alternative routes or travel times, further alleviating congestion.
- *Smart Parking Solutions*: These solutions guide drivers to available parking spaces, reducing the time spent searching for parking and contributing to a 10% reduction in vehicle emissions due to decreased idle time.

Despite its successes, Singapore faces challenges such as data privacy concerns arising from extensive real-time data collection and the high initial costs associated with upgrading infrastructure and technology. The government has implemented strict data protection regulations to address these concerns, ensuring that user privacy is prioritized.

Looking ahead, Singapore aims to expand its smart mobility solutions by integrating autonomous vehicles and enhancing data analytics for predictive traffic management. The city is also exploring the use of artificial intelligence and machine learning to further refine traffic flow and improve public transport efficiency. Public engagement initiatives will be crucial to educate residents on the benefits of ITS, fostering acceptance and encouraging greater utilization of these advanced transport systems.

Overall, Singapore's ITS implementation serves as a model for other urban areas seeking to improve their transport infrastructure and sustainability, showcasing the potential of technology to transform urban mobility.

6.2. Smart Waste Management in Barcelona, Spain

Barcelona has implemented an innovative Smart Waste Management system that leverages technology to optimize waste collection and improve urban cleanliness. The Key Features of Smart Waste Management are:

- *Smart Bins*: Equipped with sensors that monitor waste levels in real-time, these bins send data to a central management system. This allows waste collection services to schedule pickups only when bins are nearly full, rather than following a fixed schedule. This approach has led to a reduction in waste collection costs by approximately 30% and a significant decrease in the number of collection vehicles on the road, thereby lowering carbon emissions.
- *Mobile App for Residents*: The city has developed a mobile app that allows residents to report overflowing bins and track collection schedules. This app promotes community engagement by encouraging citizens to participate in waste management efforts, fostering a sense of responsibility and ownership.
- *Recycling Incentive Program*: This initiative rewards residents for proper waste separation and recycling practices. By providing incentives, the city has successfully increased recycling rates by over 20% since the program's launch.

Despite these advancements, challenges remain, including the need for ongoing public education about waste separation and the integration of new technologies into existing infrastructure. Continuous efforts are required to ensure that residents are informed about best practices and the importance of recycling.

Looking ahead, Barcelona aims to expand its smart waste management system to include more advanced analytics and artificial intelligence. These enhancements will further improve efficiency and sustainability, allowing for better resource allocation and waste management strategies.

This case study highlights how Barcelona is leveraging technology to create a more efficient and sustainable waste management system, serving as a model for other cities aiming to improve urban cleanliness and resource management.

6.3. Comparison of Intelligent Transport Systems in Singapore and Smart Waste Management in Barcelona

Both Singapore and Barcelona have implemented innovative smart city solutions to enhance urban efficiency and sustainability. Below is a comparative analysis of their approaches:

The comparative table 3 for Intelligent Transport Systems (ITS) in Singapore and Smart Waste Management in Barcelona, including more detailed aspects.

Table 3. Intelligent Transport Systems (ITS) in Singapore and Smart Waste Management in Barcelona

Aspect	Intelligent Transport Systems (ITS) in Singapore	Smart Waste Management in Barcelona
Objective	- Improve traffic flow - Reduce congestion - Enhance public transport usage - Promote sustainable transportation options	- Optimize waste collection - Increase recycling rates - Improve urban cleanliness - Foster community responsibility for waste management
Technology Used	- Smart traffic signals that adapt to real-time conditions - Real-time information systems (e.g., apps) - Electronic Road Pricing (ERP) with dynamic pricing - Smart parking solutions with guidance systems - Integrated transport apps for seamless travel	- Smart bins equipped with fill-level sensors - Mobile app for residents to report issues and track pickups - Recycling incentive programs using gamification - Data analytics for route optimization - Integration of IoT for real-time monitoring
Data Utilization	- Real-time traffic data for adaptive signal control - Historical data analysis for traffic pattern prediction - User data from transport apps for service improvement	- Sensor data for monitoring waste levels - Community reports to identify problem areas - Data analytics to optimize collection routes and schedules - Use of machine learning for predictive maintenance of bins
Impact on Efficiency	- 15% reduction in congestion during peak hours - 25% increase in public transport usage - Improved average travel times - Enhanced reliability of public transport services	- 30% reduction in waste collection costs - 20% increase in recycling rates - Reduced operational costs through optimized routes - Improved response times for waste collection

continued on following page

Table 3. Continued

Aspect	Intelligent Transport Systems (ITS) in Singapore	Smart Waste Management in Barcelona
Environmental Benefits	- 10% reduction in vehicle emissions - Decreased fuel consumption due to optimized traffic flow - Promotion of public transport reduces reliance on private vehicles	- Reduced carbon emissions from fewer collection vehicles - Increased recycling leads to less landfill waste - Improved urban air quality through reduced vehicle trips
Public Engagement	- MyTransport.SG app for real-time updates and feedback - Government-led initiatives to promote ITS benefits - Community workshops to educate on public transport options	- Mobile app for reporting overflowing bins and tracking collection - Community events to raise awareness about recycling - Incentives for residents participating in recycling programs
Challenges	- Data privacy concerns with extensive data collection - High initial infrastructure costs - Need for continuous technology upgrades - Public resistance to change in transportation habits	- Need for ongoing public education about waste separation - Integration of new technologies into existing systems - Ensuring consistent participation in recycling programs - Addressing disparities in community engagement
Future Prospects	- Integration of autonomous vehicles into the existing transport network - Enhanced data analytics for predictive traffic management - Expansion of smart mobility solutions (e.g., bike-sharing, electric vehicles) - Use of AI to optimize traffic management further	- Expansion of smart technologies to include more advanced analytics - Use of AI for waste management and predictive analytics - Development of circular economy initiatives to enhance sustainability - Collaboration with tech firms for innovative waste solutions

Sources: authors elaboration

The lessons learned from the case study of Intelligent Transport Systems (ITS) in Singapore and Smart Waste Management in Barcelona show both similarities and differences regarding urban management. The emphasis in Singapore's ITS system is on maximizing the efficiency of transportation and minimization of the congestion problem. This is done through the use of advanced technologies such as smart traffic light signals and Electronic Road Pricing (ERP). On the other hand, Barcelona concentrates on improving the Optimisation of Waste Collection and Recycling Processes through the use of smart bins and mobile applications. Both cities use data for decision-making. Singapore depend on real-time traffic data for management, while Barcelona employs sensor data for the measurement of waste. Environmental impact is of interest to both governments; Singapore attempts to lower vehicle emissions, while Barcelona attempts to reduce carbon emissions from waste collection activities. The approaches also differ in terms of the community involvement. In Singapore, the challenge has been in engaging the public by giving them real-time transport information, while in Barcelona, the approach has been to encourage active participation in waste management by reporting waste and providing incentives for recycling. Both cities have some negative perceptions regarding the lack of civic education and the introduction of new technologies, but they also have clear positive perceptions relating to the expansion of smart initiatives and the improvement of urban sustainability and efficiency.

Barcelona and Singapore showcase the benefits of smart city policies on a city's technological and data-centric capabilities. Singapore's Intelligent Transport Systems (ITS) focuses on improving the flow of traffic, minimizing congestion, and increasing the efficiency of public transport. The city is making advancements towards sustainable urban mobility through the implementation of smart traffic signals, real-time information systems, and the Electronic Road Pricing (ERP) system.

On the contrary, Barcelona's Smart Waste Management initiatives aim to improve the efficiency of waste collection and the overall recycling rate. The installation of smart bins with sensors, coupled with a mobile app for neighborhood reporting, lowered the waste collection cost by 30% while the overall recycling rate increased by 20% (Ajuntament de Barcelona, 2023). This not only improves operational efficiency but also encourages residents to partake in sustainable practices, thereby strengthening community engagement.

The innovative measures adopted by both cities emphasizes their commitment to sustainability and efficiency, which shows a potential for these urbanized regions to adopt similar strategies.

7. CONCLUSION

ITS were born from the need for traffic management and are now all linked to the quest to reduce noise and air pollution (CO2, NOx, etc.) and to optimize the impact of transport on urban and peri-urban areas. ITS reflect the need to rethink transport, both in its operation and methods, and through the paradigm through which it is analyzed. They also tend to move beyond the framework of road transport, to generalize to other areas of transport, such as maritime transport, rail transport, air transport, and in the future, space transport. Since the initial work on traffic management in the USA and Japan in the early 1970s, intelligent transportation systems (ITS) have continued to evolve, driven by technological developments and awareness of the potential they offer for improving and developing the transport sector. Thus, ITS, as an aggregating and organizing super-entity for the transportation sector, has evolved and now consists of 13 activity areas, as of 2015, as defined by the ISO-14813 standard. Furthermore, within these 13 fields of activity, it is possible to identify four major activity categories: user service, emergency services, data analysis and optimization, and finally road traffic. The fields of transportation, whether they relate to passenger transport (private or collective) or freight transport, have long been considered two separate and only loosely connected research fields. Thus, each of these fields has independently developed tools, methods, and regulations with the aim of responding to the specific problems they face. An example is the field of freight transportation, in which a multitude of optimization tools have been developed to determine the most efficient routes to ensure the delivery of goods, but which have not taken into account the congestion caused by delivery vehicles in urban areas. More recently, another example can be highlighted concerning the impact of individual mobility on collective mobility. The arrival of ride-hailing services, such as Uber, coincides with the need to facilitate mobility in urban areas, as overloaded personal and collective travel options no longer fulfill their function. Initially, a decrease in the number of vehicles in circulation was observed, but this network availability was quickly filled by a fleet of service vehicles. Subsequently, a sharp decline in public transport ridership was observed, which, in the long term, could act as a brake on the development of public transport. Noting the limitations of ad hoc responses in the transport sector, ITSs promote a systemic vision, in which as many elements as possible (whether directly or indirectly interacting) must be taken into account when optimizing a given area.

This research addresses the problems caused by traffic congestion focusing on its impacts within the context of urban mobility, economic development, and environmental sustainability. The result shows that high levels of urbanization and population density tend to generate more traffic. Such factors are worsened by the for-poor traffic infrastructure and the inefficient operation of the traffic system which causes chronically, and intermittently congested traffic. The results are severe in nature: increased time taken to travel, high consumption of fuel, and pollution of the air which affects health and contributes to climate change.

Increasing road capacity and adjusting the timing of traffic lights are common congestion relief measures that, when taken in isolation, do not provide a long-term solution. These measures attempt to treat the signs of congestion instead of the core problems. This situation makes it urgently clear that there is a growing need for thorough and flexible approaches to facilitated transport systems that would keep pace with developed urban areas.

The insertion of Intelligent Transportation Systems (ITS) seems to be an answer to these issues. Research suggests that ITS has the capability to improve the movement of vehicles and ensure safety while minimizing the effects on the environment by the use of technologies such as monitoring traffic flows in real-time.

REFERENCES

Aftabuzzaman, M. (2007). Measuring traffic congestion: A critical review. In Proceedings of the 30th Australasian Transport Research Forum (pp. 1–16). Melbourne, Australia: Engineers Australia.

Aftabuzzaman, M., Currie, G., & Sarvi, M. (2010). Evaluating the congestion relief impacts of public transport in monetary terms. *Journal of Public Transportation, 13*(1), 1–24. DOI: 10.5038/2375-0901.13.1.1

Aghabayk, K., Esmailpour, J., & Shiwakoti, N. (2021). Effects of COVID-19 on rail passengers' crowding perceptions. *Transportation Research Part A, Policy and Practice, 154*, 186–202. https://doi.org/https://doi.org/10.1016/j.tra.2021.10.011. DOI: 10.1016/j.tra.2021.10.011 PMID: 34707330

Anderson, M. L. (2014). Subways, strikes, and slowdowns: The impacts of public transit on traffic congestion. *The American Economic Review, 104*(9), 2763–2796. DOI: 10.1257/aer.104.9.2763

Bansal, P., & Graham, D. J. (2023). Congestion in cities: Can road capacity expansions provide a solution? *Transportation Research Part A, Policy and Practice, 174*, 103726. DOI: 10.1016/j.tra.2023.103726

Behruz, H., Chavoshy, A. P., Lavasani rad, A., & Mozaffari, G. (2013). Challenges of implementation of intelligent transportation systems in developing countries: Case study - Tehran. WIT Transactions on Ecology and the Environment, 179 VOLUME 2, 977–987. https://doi.org/DOI: 10.2495/SC130832

Bi, H., Ye, Z., & Zhu, H. (2022). Data-driven analysis of weather impacts on urban traffic conditions at the city level. *Urban Climate, 41*, 101065. DOI: 10.1016/j.uclim.2021.101065

Board, T. R. of Sciences Engineering, & Medicine. (2013). Deployment, Use, and Effect of Real-Time Traveler Information Systems (E. Robinson, T. Jacobs, K. Frankle, N. Serulle, & M. Pack, Eds.). The National Academies Press. https://doi.org/DOI: 10.17226/22664

Chen, X., Qi, Y., & Liu, G. (2013). Empirical Study of Gap-Acceptance Behavior of Right-Turn-on-Red Drivers on Dual Right-Turn Lanes. *Journal of Transportation Engineering, 139*(2), 173–180. DOI: 10.1061/(ASCE)TE.1943-5436.0000489

Çolak, S., Lima, A., & González, M. C. (2016). Understanding congested travel in urban areas. *Nature Communications, 7*(1), 10793. DOI: 10.1038/ncomms10793 PMID: 26978719

Falcocchio, J. C., & Levinson, H. S. (2015). *Road traffic congestion: a concise guide* (Vol. 7). Springer International Publishing. DOI: 10.1007/978-3-319-15165-6

Fan, Y., Khattak, A. J., & Shay, E. (2007). Intelligent Transportation Systems: What Do Publications and Patents Tell Us? *Journal of Intelligent Transport Systems*, *11*(2), 91–103. DOI: 10.1080/15472450701301081

Fielbaum, A., Ruiz, F., Boccardo, G., Rubio, D., Tirachini, A., & Ro, J. (2023). The job of public transport, ride-hailing and delivery drivers: Conditions during the COVID-19 pandemic and implications for a post-pandemic future. *Travel Behaviour & Society*, *31*, 63–77. DOI: 10.1016/j.tbs.2022.11.004 PMID: 36405769

Frikha, R., & Hlali, A. (2023). The Evolution Of Logistics Management Toward The Digital Transition. International Journal of Commerce. *Industry and Entrepreneurship Studies*, *3*, 66–75.

Glaister, S., & Graham, D. (2003). Transport pricing and investment in England. Imperial college (Department of civil Engineering).

Hlali, A., & Gafsi, N. (2024). Analysis of Digitalization and Sustainable Development in Africa. *Perspectives on Global Development and Technology*, *22*(5-6), 415–428. DOI: 10.1163/15691497-12341668

Jing, Q.-L., Liu, H.-Z., Yu, W.-Q., & He, X. (2022). The Impact of Public Transportation on Carbon Emissions—From the Perspective of Energy Consumption. *Sustainability (Basel)*, *14*(10), 6248. DOI: 10.3390/su14106248

Li, H., Zhao, Y., Mao, Z., Qin, Y., Xiao, Z., Feng, J., Gu, Y., Ju, W., Luo, X., & Zhang, M. (2024). Graph Neural Networks in Intelligent Transportation Systems: Advances, Applications and Trends. arXiv preprint arXiv:2401.00713. https://arxiv.org/abs/2401.00713

Lin, W., Wei, H., Yang, L., & Zhao, X. (2024). Technical review of electric vehicle charging distribution models with considering driver behaviors impacts. [English Edition]. *Journal of Traffic and Transportation Engineering*, *11*(4), 643–666. https://doi.org/https://doi.org/10.1016/j.jtte.2024.06.001. DOI: 10.1016/j.jtte.2024.06.001

Manda, V. K., Christy, V., & Hlali, A. (2025). Current trends, opportunities, and future research directions in geospatial technologies for smart cities. In Recent Trends in Geospatial AI (pp. 239–270).

Marve, S. R., Bhorkar, M., & Baitule, P. (2016). A survey on environmental impacts due to traffic congestion in peak hours. IJSTE, 2(08).

Metz, D. (2018). Tackling urban traffic congestion: The experience of London, Stockholm and Singapore. *Case Studies on Transport Policy*, *6*(4), 494–498. DOI: 10.1016/j.cstp.2018.06.002

Miles, J. C. (2014). Intelligent Transport Systems: Overview and Structure (History, Applications, and Architectures). In *Major Reference Works*. Wiley Online Library., DOI: 10.1002/9781118354179.auto166

Mohtich, F.-E., Bensiali, S., Idr, A., & El-Ayachi, M. (2023). Geospatial data integration for intelligent transportation systems: A review of the state of the art of spatio-temporal traffic forecasting. *African Journal on Land Policy and Geospatial Sciences*, *6*(4), 683–692. DOI: 10.48346/IMIST.PRSM/ajlp-gs.v6i4.41748

Rao, H., Zhang, D., Qin, G., Yue, L., & Sun, J. (2023). Investigating the intra-driver heterogeneity in car following behaviour based on large-scale naturalistic driving study. *Transportmetrica. B, Transport Dynamics*, *11*(1), 1363–1383. DOI: 10.1080/21680566.2023.2203848

Rastogi, R. (2011). Promotion of non-motorized modes as a sustainable transportation option: Policy and planning issues. *Current Science*, •••, 1340–1348.

Schrank, D., Lomax, T., & Eisele, B. (2019). 2019 Urban mobility report. Texas Transportation Institute, [ONLINE]. Available: https://Static.Tti.Tamu.Edu/Tti .Tamu. Edu/Documents/Mobility-Report-2019.Pdf

Shaygan, M., Meese, C., Li, W., Zhao, X. G., & Nejad, M. (2022). Traffic prediction using artificial intelligence: Review of recent advances and emerging opportunities. *Transportation Research Part C, Emerging Technologies*, *145*, 103921. DOI: 10.1016/j.trc.2022.103921

Silva, Ó., Cordera, R., González-González, E., & Nogués, S. (2022). Environmental impacts of autonomous vehicles: A review of the scientific literature. *The Science of the Total Environment*, *830*, 154615. https://doi.org/https://doi.org/10.1016/j .scitotenv.2022.154615. DOI: 10.1016/j.scitotenv.2022.154615 PMID: 35307440

Vijayaraghavan, V., & Rian Leevinson, J. (2020). Intelligent Traffic Management Systems for Next Generation IoV in Smart City Scenario. In Mahmood, Z. (Ed.), *Connected Vehicles in the Internet of Things: Concepts, Technologies and Frameworks for the IoV* (pp. 123–141). Springer International Publishing., DOI: 10.1007/978-3-030-36167-9_6

Wang, Y., Zhu, X., Li, L., & Wu, B. (2013). Reasons and countermeasures of traffic congestion under urban land redevelopment. *Procedia: Social and Behavioral Sciences*, *96*, 2164–2172. DOI: 10.1016/j.sbspro.2013.08.244

Yu, H., Jiang, R., He, Z., Zheng, Z., Li, L., Liu, R., & Chen, X. (2021). Automated vehicle-involved traffic flow studies: A survey of assumptions, models, speculations, and perspectives. *Transportation Research Part C, Emerging Technologies, 127*, 103101. https://doi.org/https://doi.org/10.1016/j.trc.2021.103101. DOI: 10.1016/j.trc.2021.103101

Zemmouchi-Ghomari, L. (2025). Artificial intelligence in intelligent transportation systems. Journal of Intelligent Manufacturing and Special Equipment. https://doi.org/DOI: 10.1108/JIMSE-11-2024-0035

Zhang, S., Li, J., Shi, L., Ding, M., Nguyen, D. C., Tan, W., Weng, J., & Han, Z. (2023). Federated Learning in Intelligent Transportation Systems: Recent Applications and Open Problems. arXiv:2309.11039. https://arxiv.org/abs/2309.11039

Zheng, Z., Wang, Z., Zhu, L., & Jiang, H. (2020). Determinants of the congestion caused by a traffic accident in urban road networks. *Accident; Analysis and Prevention, 136*, 105327. DOI: 10.1016/j.aap.2019.105327 PMID: 31874332

Chapter 9
The Problematic Field of Implementing Technological Innovations in Mobility Systems on the Path to Sustainable Development

Natalia S. Alekseeva
https://orcid.org/0000-0002-7355-3277
Peter the Great St. Petersburg Polytechnic University, Russia

Svetlana V. Pupentsova
https://orcid.org/0000-0003-3742-0482
Peter the Great St. Petersburg Polytechnic University, Russia

Maria G. Livintsova
https://orcid.org/0000-0003-3565-3880
Peter the Great St. Petersburg Polytechnic University, Russia

Habib Noorbhai
University of Johannesburg, South Africa

Olga V. Kalinina
https://orcid.org/0000-0002-4617-2823
Peter the Great St. Petersburg Polytechnic University, Russia

DOI: 10.4018/979-8-3373-0882-1.ch009

ABSTRACT

Mobility systems are now at the center of social, economic and environmental agendas, influencing quality of life, spatial accessibility and carbon footprint reduction. Nevertheless, innovation faces a number of challenges such as resource scarcity, knowledge lacunas and conflicting interests of parties. The aim of the study is to analyze the challenges of implementing technological innovation of mobility systems towards sustainable development. In this paper, mobility systems are considered as complex socio-technical systems, which creates the need for interdisciplinary research to study the mentioned problems of technology innovation adoption. The significance of the research results lies in understanding the problem field of implementing technological innovation in urban mobility systems for sustainable development. The results provide a basis for developing strategies to overcome its barriers - resource scarcity, social and economic inequalities, and conflicts of stakeholder interests.

INTRODUCTION

The current global situation calls for a transition to sustainable development, which has become one of the key challenges for society. Despite the existence of well-established models, approaches, and practices for achieving this transition, pressing issues remain, such as the need to scale these solutions and adapt them to modern challenges and new conditions (Management Association, 1 C.E.). The essence of the transition lies in the necessity for society to revise the existing social contract. Kalshoven and Zonderland (Huntjens, 2021) suggest that the social contract is an implicit agreement among members of society to create a better and safer shared life. This contract outlines how public goods and services are distributed based on the balance of contributions and benefits.

New challenges in the socio-political agenda have become catalysts for the necessity to overhaul this contract, including changes to international models of relations between governments and economies (Vuković et al., 2020), between economies and civil society, as well as among commercial entities (Atiku et al., 1 C.E.). Today's global-scale problems underscore the need for a systemic shift from addressing localized issues to tackling broader national and international interests (Popescu et al., 1 C.E.). As a practical solution to these global challenges, the United Nations adheres to its 17 Sustainable Development Goals, which provide a roadmap for governments, businesses, and civil society. Achieving this roadmap would allow for a balance among the economic, social, and environmental priorities of the world.

One of the ways to solve the problems of sustainability is the development of innovative transport technologies (Bril et al., 2016; Didenko et al., 2021; Kalinina et al., 2022). Mobility systems are now at the center of social, economic and environmental agendas, affecting quality of life, territorial accessibility and reducing the carbon footprint. Nevertheless, innovations face a number of challenges, such as lack of resources, knowledge gaps and differences of interests between the parties (Babkin et al., 2022). The problematic field of technological innovations of mobility systems includes both technical and socio-economic aspects, which requires an interdisciplinary approach and a comprehensive solution (Dubolazov et al., 2022).

For a successful transition to sustainable social development, it is necessary to ensure that all costs, benefits and risks are distributed fairly among all actors: citizens, business and the public sector (Fadeev et al., 2024). Only joint efforts will allow us to obtain innovative solutions that help achieve the ideal of sustainable development. Therefore, the studied field of problems of sustainable participation of technological innovations in mobility systems will reveal the challenges of creating the necessary conditions for a fair, inclusive and environmentally safe transition. The transition poses a threat to the existing practice, politics and philosophy of the world order, but it offers the development of new knowledge about the ways of developing a new world order. Therefore, knowledge of the main problems and opportunities will help to develop the most effective approach to maintaining the accelerated deployment of development programs that comply with the principles of sustainability.

The goal of this study is to analyze the challenges of implementing technological innovations in mobility systems on the path to sustainable development.

MATERIALS AND METHODS

This study analyzes the problematic field of the introduction of technological innovations of mobility systems for sustainable development using a mixed methodological approach. The emphasis is on qualitative methods of studying the technical, socio-economic and environmental aspects of the problem field. We conducted a systematic review of the literature on sustainable mobility and technological innovation and studied various types of research analytics. Scientific articles, corporate reports and policy documents were selected as primary sources in order to ensure the widest possible coverage of existing knowledge on the research topic. In addition, we used the Case Study method to study practical examples of the application of innovations in the field of mobility. These studies have provided a better understanding of the benchmarks of industry and corporate solutions.

RESULTS

In recent years, the global community has increasingly adhered to the principles of sustainable development, which have become an integral part of both society and the socio-economic advancement of nations. Sustainable development, in this context, refers to the gradual transformation of economic systems—whether national economies or individual organizations—that aims to meet the needs of the present while consistently progressing toward improvement or achieving a socio-economic balance between internal and external environments. This process relies on the efficient use of resources, the preservation of the natural environment, and the long-term enhancement of organizational value. Sustainable development is characterized by a balanced approach that incorporates environmental, social, and corporate governance (ESG) factors—where "E" stands for environmental, "S" for social, and "G" for governance.

In Russia, support for the ESG agenda can be observed at all levels, as its basic principles are in line with Russian values for the development of both the state and international relations with new strategic partners.

Modern global changes call for revision of many business processes through the prism of ESG agenda. The purpose of the revision is to improve the quality of life of the society, to reduce the risks that arise in the implementation of investment policy at the state level. All this is important to ensure the strategic development of the country. Russia is considered a country with great prospects and in modern conditions follows the world trends towards sustainable development.

In the strategic development of the country's economic entities the importance of ESG-factors assessment supported at different levels is increasing (Figure 1).

Figure 1. Levels of support for ESG-transformation

Conceptual level (neo-communitarianism and stakeholder capitalism)

- Corporate Social Responsibility (CSR) – social responsibility concepts
- **The concept of sustainable development based on the principles of** Environmental, Social, and Corporate Governance (ESG):
- the principle of responsibility towards the environment. It consists in the efficient use of resources and in the formation of the organization's environmental policy. Policies aimed at minimizing or eliminating harmful impacts on airspace and natural resources, efficient waste management and promotion of biodegradable products, efficient use of energy resources (E-principle);
- the principle of responsibility towards the strategic development of the country and society. It includes support of social programs, social responsibility (business ethics) to employees and other contractors of the enterprise, training and development of human resources, ensuring employment of the population, as well as participation in a variety of charitable organizations and foundations, conducting charitable actions with minimal government intervention in private business (S-principle);
- the principle of supporting the quality of management at a high level. It includes effective functioning of the management system and risk management. For its realization it is necessary to have a well-developed strategy of the company's development in a constantly changing external environment. The principle sets a strategic focus on increasing the value of equity capital and disclosure of corporate reports required for ESG rating (G-principle).

Normative level

- Sustainable Development Goals
- United Nations or a set of 17 interconnected goals developed in 2015 by the UN General Assembly as a "blueprint for achieving a better and more sustainable future for all".
- The Paris Climate Agreement was adopted on December 12, 2015, following the 21st Conference of the Framework Convention on Climate Change in Paris.
- **Cross-country and national regulations:**
- Strategy of socio-economic development of the Russian Federation with low greenhouse gas emissions until 2050;
- Federal Greenhouse Gas Emissions Control Act of 2021;
- Federal Law of 2022 "On Conducting an Experiment to Limit Greenhouse Gas Emissions in Certain Constituent Entities of the Russian Federation";
- Order of the Russian Government of 2021 "On Approval of the List of Greenhouse Gases for Which State Accounting of Greenhouse Gas Emissions and Maintenance of the Greenhouse Gas Inventory is Carried Out" Legal acts on taxonomy and project support measures: Order of the Russian Government of 2021 "On Approval of the Goals and Main Directions of Sustainable Development of the Russian Federation";
- Russian Government Resolution of 2021 "On Approval of Criteria for Sustainable Development Projects in the Russian Federation and Requirements for the Verification System of Sustainable Development Projects in the Russian Federation";
- Decree of the Bank of Russia dated 2021 "On Standards of Securities Issuance", etc.

Reference level (evaluation and monitoring)

- Standards (more than 70 standards, the best known being CDP, CDSB, GRI, SASB, IIRC and TCFD)
- ratings (more than 600, including national ratings)

Project level

- ESG initiatives of specific countries, companies and regions

(compiled by the authors)

How important and necessary ESG-transformation is stated in many development programs of economic entities of the world community.

Urban mobility serves as the foundation of sustainable territorial development, influencing factors such as economic growth, social inclusion, and environmental protection. Rapid urbanization and ecological crises have amplified the need to reassess traditional mobility systems to integrate mobility innovations (Pupentsova et al., 2022). This study examines the problematic field and challenges, focusing on issues such as limited resource access, socio-economic inequality, and stakeholder interest conflicts.

In 2024, 150 billion rubles have been allocated in the Russian Federation for infrastructure development, construction of electric bus fleets and increasing the efficiency of urban mobility. In Moscow in the next 10-15 years, strategic preference is given to the development of the metro and the Moscow Central Diameter. In St. Petersburg in 2025, the main focus is on the construction of new metro stations, improvement of public transportation service, and infrastructure reconstruction to expand the segment of urban electric trains. To date, the St. Petersburg metro covers only 30% of the city.

It should be noted that with a population of 13 million people in Moscow, 393 stations have been built, which is 33,460 people per 1 station. Meanwhile, in St. Petersburg, with a population of 5.6 million people, there are only 72 stations, i.e. 77,777 people per station. The overloaded stations of the second capital of the Russian Federation push local authorities to pay increased interest to this issue.

The main technological breakthroughs in the field of urban mobility in Moscow and St. Petersburg include:

1) increase in the number of electric vehicles (electric supercars) and large-scale expansion of infrastructure to serve this type of transport in urban environments, which contributes to environmental protection and increases the efficiency of freight transportation;
2) equipping vehicles with automated safety systems, computer vision, navigation system to reduce the risks of emergency events on the roads, reducing risks arising from uncontrollable human factors;
3) the development of a new system for the transportation of goods and services;
4) expansion of the industry for recycling and disposal of decommissioned vehicles and electric vehicle batteries, which fully contributes to improving waste management efficiency;
5) introduction of unmanned technologies in service (introduction of delivery drones) and moving mini-warehouses of goods to the final consumer to reduce transportation congestion;
6) development of ecological systems in the logistics direction, which reduces routes and time spent in traffic jams;
7) Increasing the number of Internet of Things sensors embedded in vehicles to monitor equipment health and performance.

The following can be attributed to the promising directions of development of technological innovations of mobility systems on the way to sustainable development.

1. Increase in the number of electric vehicles and large-scale expansion of infrastructure to serve environmentally friendly transportation in urban environments.
2. Equipping vehicles with automated safety systems.
3. Use of ecological materials in the production of vehicles.
4. Expansion of the industry for recycling and utilization of decommissioned vehicles and electric vehicle batteries.
5. Expansion of the industry for recycling and utilization of decommissioned vehicles and electric vehicle batteries.

6. Development of ecological systems in the logistic direction to reduce routes and time spent in traffic jams.
7. Increasing the number of embedded Internet of Things sensors in vehicles to monitor the condition and performance of equipment.

Let's look at these areas in more detail.

1. Increase in the number of electric vehicles and large-scale expansion of infrastructure to serve environmentally friendly transportation in urban environments

The global electric vehicle market is expected to grow to 16.7 million units in India, Thailand and Indonesia in 2023. There was a sharp increase in sales of electric vehicles. Sales of electric vehicles in the "Rest of the World" category are expected to reach 840,000. As of January 1, 2024, 39,056 electric vehicles were registered in Russia, of which 63.5% are registered in 10 regions (Figure 2).

So far, only electric buses are produced in Russia. But in accordance with the "Concept for the development of production and use of electric vehicles in the Russian Federation until 2030," approved by the Russian Government in 2021, about 220 thousand electric cars should be produced annually in Russia by 2030. The necessary infrastructure will be created for them, and part of it is factories for recycling used batteries.

Figure 2. Electric vehicle fleet by district and region, pcs.

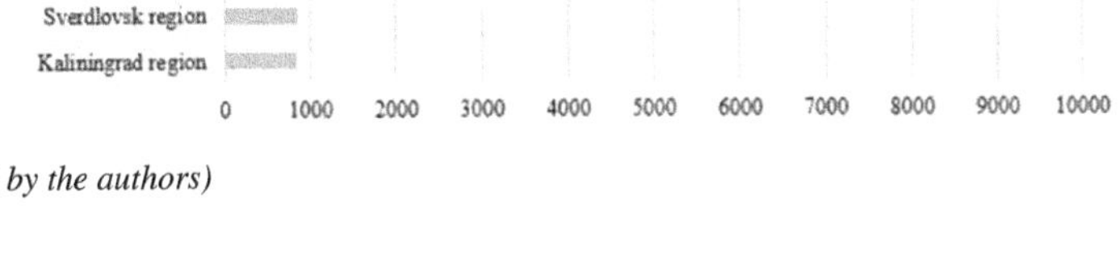

(compiled by the authors)

In 2024, the number of charging stations for electric vehicles in Russia increased by almost 70% and reached 7,410 units, including 2,555 units with a capacity of 50 kW or more. According to the concept for the development of mobility systems on the path to sustainable development, by 2030 we expect an increase in charging stations for electric vehicles at the level of 72 thousand units and about 1.5 million electric vehicles. Investments in the construction of ultrafast charging stations for electric vehicles are estimated at 5-6 million rubles. Experts note that the charging station pays off quickly if it is loaded by at least 30%.

2. Equipping vehicles with automated safety systems

Automated modern systems for equipping vehicles and related infrastructure include continuous monitoring and video surveillance on highways; traffic management, public transport on urban highways, in emergency situations, and in the aftermath of accidents; automatic collision avoidance and safety of vehicles; management of large information flows; combined freight transportation; weather monitoring on highways; digital payment systems for transport services; maintenance and operation of highways; navigation information for road users (Lukashevich et al., 2021; W. Zhang et al., 2023).

3. Use of ecological materials in the production of vehicles

In modern models, attention is increasingly being paid to the interior trim. In cars, you can see synthetic textiles instead of leather upholstery. This synthetic textile is obtained by recycling plastic bottles. There is a trend of replacing toxic products with eco-friendly bioplastics from recycled materials and materials based on natural flax and eucalyptus fibers. Toyota uses sugar cane glycol to produce seat linings in its cars, while Ford uses soy bean foam. Many companies offer a wide range of environmentally friendly car care materials. In this regard, EPP foam polypropylene is increasingly being used in the automotive industry. The flexibility of this material makes it possible to create automotive components of various shapes. For example, due to the increased impact resistance, EPP elements are ideally suited as shock-absorbing parts, and their low weight increases the efficiency of movement. Due to the possibility of complete recycling of this material, the automotive industry is moving towards "green" technologies, thereby reducing its impact on the environment.

In the manufacture of cars, they are increasingly switching to closed circuits through which water circulates, which is used for processing plastics. This principle significantly reduces water consumption and is more environmentally friendly. Also, in the process cycles, we also use condensate from the cooling and heating systems.

4. Expansion of the industry for recycling and utilization of decommissioned vehicles and electric vehicle batteries

In 2018, the national Ecology project was launched in Russia. Its budget amounted to 4,041 billion rubles and was allocated to 11 federal projects. The projects selected were measures to improve the health of water bodies, reduce emissions of hazardous pollutants, and create a sustainable solid municipal waste management system. The national objectives of the project are presented in Decrees N° 204 and 474 of the President of Russia (Figure 3).

For the successful implementation of the environmental strategy, it is important to involve all stakeholders: the public, business and the government. It should be noted that the leading roles in this process are assigned to the state and large business (Figure 3). These major players should choose development guidelines for the implementation of long-term projects, incentives and control.

Figure 3. National goals and objectives in the decrees of the President of the Russian Federation

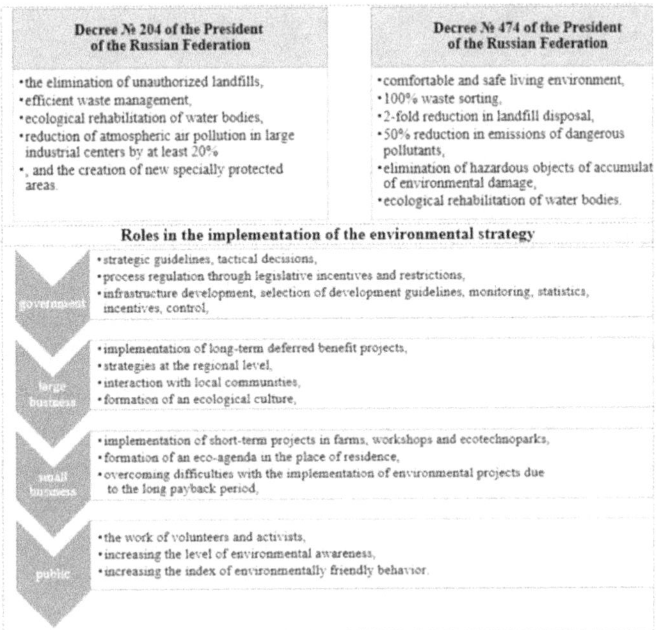

(compiled by the authors)

The solid waste management system has been reformed in the Ecology project. In recent years, the processing of waste of hazard classes I and II has been transferred to the federal operator, a unified state information system for accounting and control over the management of such waste has been introduced and built. The indicator for the elimination of the most dangerous facilities (abandoned industrial enterprises and landfills) has been completed.

The federal Clean Air project has approved a comprehensive plan of measures to reduce emissions, and provides preferential loans for projects on the disposal of pollutants and their purification.

The main indicators of the national Ecology project are shown in Figure 4.

Figure 4. Main indicators of the national Ecology project

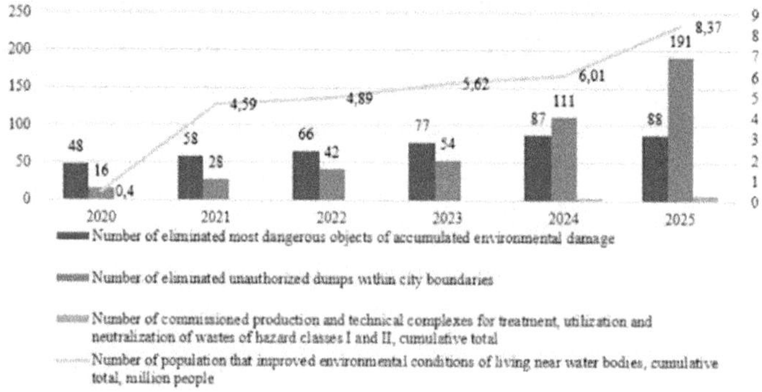

(developed by the author based on the sources used: Ministry of Natural Resources and Ecology of the Russian Federation)

Decommissioned batteries pose a particular danger to the environment, which are prohibited from being simply disposed of in landfills. A standard electric car battery with a mileage of about 20 thousand km per year will last about 15 years. According to global forecasts, more than 3.4 million battery packs of electric vehicles will need to be replaced by 2025. Lithium-ion batteries are recycled all over the world by small laboratories, as well as industrial enterprises. At the moment, there are at least hundreds of them. But still, automakers that create separate factories or workshops for their batteries show the greatest efficiency. Many companies independently solve the problem of recycling decommissioned batteries. There are 37 kg of chemically hazardous electrolyte, 126 kg of aluminum and a total of 102 kg of nickel and graphite per 400 kg of battery. And due to the reuse of rare metals, the cost of batteries is reduced by half.

The batteries of "green" cars are recycled using the same technology as lithium-ion batteries for smartphones, laptops and tablets. To do this, it is enough to chop the battery, and then divide it into its component elements. The main problem with electric car batteries is that they are huge. Before you start chopping, you need to separate the case from the batteries, as well as disconnect the electronic components that can be reused. Then the production needs an oxygen-free shredder to prevent fire, or special protection to keep the process under control. At the exit of the shredder, a crushed mass is obtained, which must be divided into components, which is an even more time-consuming procedure. Therefore, the fission process and the amount of mixtures coming out depend on the technologies used at the plant and are kept secret. Most often, copper and aluminum are separated into one mixture, while rare metals such as cobalt, nickel, lithium and manganese go into another group. Plastic closes the sorting.

The attitude of the population towards waste recycling is also important in addressing waste disposal issues. According to the Romir study, the attitude of Russians towards the environment has changed in a positive way over the past 25 years: the choice between environmental protection and economic growth has changed from 46/30 to 36/50, and the proportion of respondents who are not interested in this problem has decreased from 24 to 13% (Figure 5).

Figure 5. Russians' choice between environmental protection and economic growth

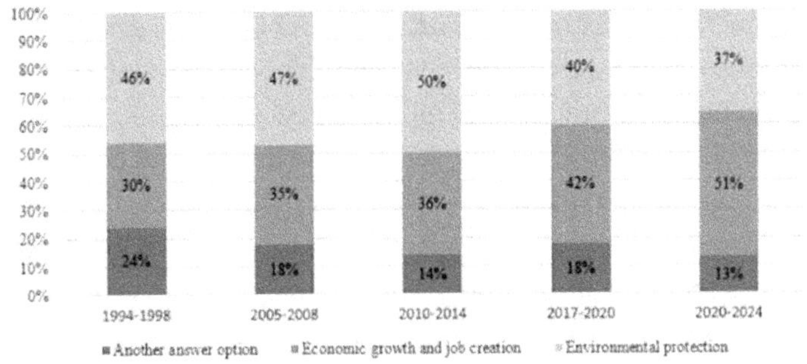

(developed by the author)

Thus, the urgency of environmental issues is increasingly attracting the attention of the Russian public. Many Russians strive to contribute to the protection of the environment. The environmental reform started with decent results despite the geopolitical and economic difficulties. It has affected all sectors of society and business, and continues to gain momentum towards achieving the 2030 environmental targets.

Nevertheless, among the barriers to environmental awareness of the population, there are:

- Abandoning eco-friendly services and products in favour of cheaper but less environmentally friendly options;
- low awareness not only about environmental problems and ways to overcome them, but also about everyone's contribution to solving environmental problems;
- weak government support for businesses and the public.

The presented environmental agenda of the national development goals of Russia until 2030 provides grounds for further changes in the external environment of organizations. Achieving these goals will provide a powerful lever for reviewing the methods of evaluating the effectiveness of an organization's management, taking into account the long-term goals of sustainable development and environmental protection.

5. Expansion of the industry for recycling and utilization of decommissioned vehicles and electric vehicle batteries

The use of unmanned vehicles in business processes has increased dramatically over the past five years. It is mainly used to monitor the condition of power lines, to create 3D models, digital doubles, to spray fields with pesticides and fertilizers. According to analysts, investments in drone production will grow to $47.76 billion by 2025. Therefore, in the near future, we can expect a breakthrough in the use of unmanned vehicles in logistics and cargo delivery.

According to Gartner, by 2026, about 1 million drones will be used worldwide to deliver goods to consumers. An analysis of world practice shows that so far, unmanned vehicles are mainly used for transporting light goods. Large-scale territories of Russia can become an excellent testing ground for experiments in this field. Pilot projects for the use of drones are already being implemented in the field of e-commerce, as well as the transportation of postal parcels. The national technological initiative was launched by the scientific and production enterprise New Telecommunications Technologies, using unmanned vehicles at the Vostochno-Messoyakhskoye field and delivering a total of 150 kg of cargo to the field.

6. Development of ecological systems in the logistic direction to reduce routes and time spent in traffic jams

This category includes the optimization of delivery routes, which reduces the distance traveled and depreciation of vehicles, thereby reducing the environmental burden.

The introduction of warehouse management systems (WMS) and transportation (TMS) reduces the time of cargo delivery and allows you to adjust the monitoring of vehicle movement and driving style of drivers, including through satellite monitoring. The introduction of artificial intelligence and machine learning systems, as well as the use of big data processing and analytics results, contributes to more accurate forecasting and automation of processes.

Digital management systems simplify the interaction between suppliers and carriers, making processes more transparent and faster. In such a situation, the competitiveness of logistics operators increases and the entire logistics industry develops.

Under the influence of sanctions, Russia has developed a trend away from foreign software. The political situation contributed to the active growth of the national software market in 2024. New Russian logistics management systems have emerged, characterized by increased reliability and data security, as they take into account the specifics of the Russian logistics infrastructure and legislation. Here we note the ecosystem of ITOB logistics solutions that allow operators to effectively manage supply chains and transportation of goods. 1C:TMS Logistics and 1C:Satellite Monitoring Center products are offered for route planning and cargo control in the domestic market. These software products simplify interaction with shopping malls and delivery companies. All programs are of national origin, perfectly integrated with international ERP systems.

7. Increasing the number of embedded Internet of Things sensors in vehicles to monitor the condition and performance of equipment

The concept of the Internet of Things (IoT) arose as a result of the large-scale connection of various devices to the Internet. The official definition of the Internet of Things is proposed by the ITU-T Y.2060 "Overview of the Internet of Things" Recommendations. The concept of IoT is formulated as "a global information society infrastructure that provides advanced services by organizing communication between things (physical or virtual) based on existing and developing compatible information and communication technologies" (Arbabzadeh & Kermani, 2019).

The emergence of the IoT has become a new stage in the evolutionary development of the Internet (Brouwer et al., 2018). IoT is an interaction using heterogeneous networks and multiple sensors controlled by a single standard. The task of the infocommunication industry is to create the value of IoT for every person. With the development of the IoT market, the cost-effectiveness factor in the introduction of new "smart" technologies is becoming more and more understandable and obvious not only for consumers, but also for logistics companies. Transport companies are attracted by the opportunity to increase resource efficiency through IoT solutions and services (Huang et al., 2020). Automotive IoT integrates Internet of Things (IoT) technologies into the automotive industry by connecting cars to the Internet and enabling communication between cars, drivers, and various external systems. This integration allows the exchange of data and information, providing new functions and services and increasing overall efficiency and security (Y. Zhang et al., 2020).

Our research has revealed the following problems, which are comprehensively related to each other (Figure 6).

Figure 6. The problematic field of introducing technological innovations in mobility systems

Missing image: \\IGI-vFS01\Production_Image\Books\2026\9798337308821\979-8-3373-0882-1.ch009.fa.tif (979-8-3373-0882-1.ch009.fa).

(compiled by the authors)

Let's look at each point in more detail.

1. Technological capabilities

In the transportation industry, about 80% of companies are partially or completely switching to new business models based on digital technologies. The industry has already created and is actively implementing digital platforms in the field of passenger ticket sales (for example, aviasales.ru), freight transportation (for example, Russian Railways Freight Transportation), taxi aggregators (for example, Yandex. Taxi) and others.

Digitalization of various sectors of the transport industry is taking place unevenly.

The greatest progress is being made in areas such as customer engagement (including through digital channels, the development of aggregator platforms) and change management. The creation of digital business models, the introduction of digital platforms and flexible development methods within organizations demonstrate average levels of implementation.

The lagging areas are the digitalization of support functions and operations within the organization. There is a shortage of personnel with the necessary digital skills. The main barriers to the digitalization of the industry are the lack of specialized personnel and financial resources, inefficient standards and regulations, including in terms of electronic document management and the provision of public services, and the low priority of digital transformation for many participants in the transport market.

According to the passport of the "Strategy of Digital Transformation of the transport industry of the Russian Federation" updated for 2023, active development of the information technology market is planned. By 2030, it is expected:

- the growth of domestic logistics ecosystems by 4 times compared to 2019;
- transformation of the urban environment and qualitative transformation by 50%;
- socialization of the basic services of the transport industry, available electronically up to 95%;

- active implementation of AI in priority projects related to the introduction of unmanned vehicles and the introduction of biometrics for passenger admission (it is expected that barrier-free fare will be used on bus routes in 600 cities by 2030);
- the growth of specialists who intensively use information and communication technologies to 10.8 million by 2030.

The table below provides a list of the most reputable information technology providers in the transport and logistics industry.

Table 1. The most reputable information technology providers in the transport and logistics industry

Organization	Revenue from projects in the transport industry in 2023, million rubles	Revenue growth ratio relative to 2022
Urbantech	11 349	28%
Glonass	4 954	127%
Korus Consulting	4 411	74%
Software product	4 122	28%
Smart Driving Lab	3 671	739%

(compiled by the authors)

It should be noted that in the near future, Russia plans to create a National Digital Transport and Logistics Platform (NCTLP), which will include government information systems, digital services and a library of regulatory documents used in the organization of transportation. Thanks to the platform, the average speed of goods delivery is expected to increase to 15%, and the volume of cargo transportation will grow to 12%.

The state information system of electronic transportation documents (GIS EPD), launched in the fall of 2022, will become part of the national transport platform. The Federal Tax Service, the Ministry of Finance and the Ministry of Internal Affairs of Russia participated in its development. In 2023, using GIS EPD, it was possible to electronically issue 6 types of documents, including waybills and waybills.

A pilot project on electronic registration of multimodal transportation has been launched with the participation of Russian Railways, FESCO and the Digital Transport and Logistics Association. During the project, the possibility of ordering cargo delivery in a single window mode involving different types of transport – automobile, rail, water and air transportation is being worked out. At the first stage, the road–rail transport chain is being tested. At the next stage, the connection of another link, marine transport, will be worked out on the basis of the Vladivostok Commercial Sea Port.

Two more national projects will be launched in the near future:

- The national project "Efficient Transport System" will include eight federal projects in the areas of transport. As part of the federal digitalization project, it is planned to create a state information system for transport and economic balance. This system will allow you to identify infrastructure barrier sites, analyze the actual and projected volumes of cargo and passenger traffic. In turn, the National Digital Transport and Logistics Platform being developed will be aimed at removing excessive administrative barriers for market participants, as well as increasing the efficiency of using transport infrastructure.
- The national project "Infrastructure for Life" will include six federal projects related to the road sector. In particular, it provides for the use of digital technologies to reduce the level of accidents and provide citizens with public transport on popular routes.

Thus, the government sees the digitalization of transport as the most important tool for overcoming logistical problems and actively supports this direction. Today, Russian IT companies are facing a serious challenge - they will have to create full–fledged digital tools for the transport and logistics industry as soon as possible.

2. Barriers to digital literacy

The main barriers to digital literacy of employees in the transport and logistics industry are related to staff resistance. The reason for the resistance of employees of enterprises may be the reluctance to switch to new systems. Because these solutions make the logistics processes in the company more transparent, but not everyone is ready to show this transparency.

In the context of modern challenges and changes, it is important to apply a combined approach to training, combining e-courses and conferences with seminars, in order for employees to master the necessary digital skills.

It is proposed to invite top management of companies as mentors and mentors, this will make it possible to achieve great success in retraining staff. With the advent of information and communication technologies such as videoconferencing, distance learning platforms, social networks and online communities, and thematic Internet portals, remote communication prevails over face-to-face communication. With the advent of digital technology, a new type of reverse mentoring has emerged. Reverse mentoring is expressed in the fact that a young professional becomes a mentor to an experienced employee on new trends and technologies. At the same time, an experienced specialist becomes a mentor to a young employee in matters of methodology and organization of work and educational processes. Having mastered the capabilities of the digital ecosystem, employees will have the opportunity to work remotely without loss of efficiency, and managers will gain an assistant in the form of operational, unbiased analytics. Among modern approaches to mentoring in an organization, individual and group communication, direct (through consultations and classes) and indirect (through advice and recommendations) are distinguished.

In the context of constant changes and the active introduction of digital and innovative technologies, agile mentoring is developing in a continuous spirit. Agile mentoring cannot be guided by clearly defined work standards, as they are developed in the course of the work itself. Today, agile mentoring is becoming more common, carried out within the framework of project cross-functional teams without hierarchical relationships. Project teams are both creators and bearers of experience, and their members provide mutual mentoring.

To increase the interest of employees in mastering digital literacy, the introduction of digital skills indicators in KPIs not only for ordinary employees, but also for top managers.

To simplify the stage of customer adaptation to a new IT solution, the developers suggest following the recommendations for effective launch. Experts advise adding all flights to the site, motivating logistics specialists and at the same time monitoring their work, setting an order price commensurate with market conditions, as well as notifying carriers about the transition to a modern ecosystem.

3. Equity in accessibility

Modern technological innovations in mobility systems, such as autonomous vehicles, intelligent transportation systems and sharing platforms, have the potential to transform the urban landscape and improve the quality of life. However, equity in access to these technologies remains an important and under-researched challenge. Socially vulnerable groups - including the elderly, people with disabilities, low-income populations and people living in rural areas - are often excluded from the adoption and use of innovative solutions. Reasons include both physical inaccessibility of infrastructure and the digital divide, lack of information, as well as a lack of consideration of diverse needs at the design stages.

Achieving true inclusivity requires a systemic rethinking of mobile technology design and deployment processes. This involves involving representatives of different communities in decision-making, adapting interfaces and services to a variety of cognitive and physical abilities, and government support for initiatives to remove barriers. Equity in technology accessibility is a key prerequisite for equitable and sustainable urban mobility, requiring an interdisciplinary approach and a long-term strategy.

4. Cybersecurity

With the development of digitalization, the transport and logistics industry is becoming increasingly vulnerable to cyber threats. The adoption of intelligent transportation systems, the Internet of Things (IoT), autonomous vehicles and cloud-based logistics management platforms increases the number of potential entry points for cyberattacks. Disruptions to these systems can lead not only to financial losses, but also to major supply chain disruptions, transportation collapses and threats to human safety.

One of the key challenges is the lack of cyber resilience of the technologies in use. Many transportation companies are still running outdated software, do not regularly update security systems, and lack effective incident response procedures. In addition, there is a shortage of cybersecurity specialists capable of tailoring protection to the specifics of logistics processes.

To minimize risks, it is necessary to integrate cybersecurity into industry strategic planning, develop industry standards and conduct regular audits. Special attention should be paid to protecting personal and commercially sensitive data, as well as ensuring resilience to attacks on critical infrastructure.

Thus, cybersecurity is becoming an integral component of sustainable and secure functioning of transportation and logistics systems in the context of digital transformation.

5. Structural shifts in the labor market

Automation and digitalization are reducing demand for traditional professions (drivers, logisticians) while creating a need for new competencies (AI developers, big data analysts). These changes are accompanied by an increase in structural unemployment due to the mismatch of workers' skills with the requirements of the digital economy.

The key problem is the uneven adaptation of labor resources: low-skilled workers face the risk of displacement, while highly skilled workers benefit. In addition, the geographic concentration of high-tech jobs exacerbates regional disparities. Balanced policies will minimize social costs and ensure an inclusive transition to the economy of the future.

6. Attracting financing

The creation of sustainable and equitable financing models is an important prerequisite for the development of innovative mobility systems that can ensure the adaptation of cities to the challenges of the future. Such innovative projects often face a high level of technological and market uncertainty, which reduces their investment attractiveness. Financing can come from various sources: government budgets, venture funds, private investments, international grants, and public-private partnerships. However, there is an imbalance between the number of promising projects and the available funds, especially in developing countries and small towns. To increase the effectiveness of financing, it is necessary to develop mechanisms for assessing risks and socio-economic effects, introduce flexible financing models, and encourage private sector participation through tax incentives and guarantees.

7. Cross-border issues

Cross-border issues in the transport and logistics industry are an important factor affecting the efficiency of global supply chains. The growth of international trade, the development of electronic markets and the deepening of economic integration reinforce the need for coordinated logistics between countries. However, differences in customs procedures, regulatory requirements, security standards, as well as in the level of digitalization and infrastructural development create serious barriers to the smooth movement of goods.

One of the key problems remains the lack of harmonization of regulations and procedures between countries. This leads to delays at borders, increased costs and risks, especially in multimodal transport. In addition, political instability, sanctions regimes, and restrictive measures imposed in emergency situations (such as pandemics) can seriously disrupt the stability of cross-border logistics flows.

To overcome these challenges, an integrated approach is needed: the development of international agreements, the introduction of digital solutions (for example, electronic waybills and digital customs), as well as strengthening institutional cooperation between countries. An important area is to increase transparency and traceability of logistics operations, which helps reduce corruption risks and increase trust between market participants.

8. Problems of long-term cooperation

Problematic is the need to harmonize the interests of various participants in the logistics chain, the complexity of forecasting and risk management while ensuring the quality and safety of transport services. In the context of long-term development, it is necessary to comply with international standards, as well as to develop and implement effective economic mechanisms for coordination and risk management.

This may include the creation of specialized platforms for information exchange among supply chain actors, the development of quality and safety standards, and the implementation of monitoring and control systems. In addition, it is important to take into account cultural and linguistic differences between actors involved in international cooperation. This can make it difficult to communicate and understand goals and objectives. It is therefore necessary to develop and utilize tools to overcome these barriers, such as translators and glossaries of terms.

DISCUSSION

The findings of this study align with the broader scientific discourse on sustainable development challenges. (Verbong & Geels, 2010) emphasized that sociotechnical systems require transformation to support sustainable development (*Automobility in Transition?*, n.d.). Our findings confirm the existence of key barriers identified by Geels, such as high technological costs and regulatory complexities, which are similarly noted in Barter's (2019) analysis of transport system adaptation under resource constraints (Barter & Russell, 2012).

This study's results are particularly aligned with research on mobility innovations by (Cohen & Kietzmann, 2014), who emphasized the role of engaging local stakeholders in successfully adopting mobility innovations.

The theoretical framework of this study on the problematic field of technological innovations in mobility systems is grounded in the concept of sustainable development and the theory of sociotechnical transitions. Mobility systems are considered complex sociotechnical systems, necessitating interdisciplinary research to address the challenges of implementing innovative technologies. These challenges include a lack of financial resources and equipment, insufficient knowledge about adapting innovations to local conditions, and power struggles among stakeholders. It is crucial to align the interests of the state (Vashkevich et al., 2024), the private sector, and civil society, as well as to enhance digital literacy and public involvement in technology adoption processes.

CONCLUSIONS

The significance of this research lies in understanding the problematic field of implementing technological innovations in urban mobility systems for sustainable development. The findings form a foundation for developing strategies to overcome barriers such as resource scarcity, social and economic inequality, and stakeholder interest conflicts.

The analysis indicates that these innovations require not only technical adaptation but also changes in institutional structures, regulatory frameworks, and behavioral models. Our results further highlight that various technologies—including electric vehicles, shared mobility platforms, and autonomous systems—should be integrated into multimodal networks to reduce the carbon footprint, promote territorial equity, and improve urban environments.

REFERENCES

Arbabzadeh, M., & Kermani, M. J. (2019). Battery Recycling and Its Effect on Environment. In *Handbook of Materials Failure Analysis with Case Studies from the Aerospace and Automotive Industries* (pp. 561–575). Butterworth-Heinemann.
Atiku, S. O., Jeremiah, A., Semente, E., & Boateng, F. (1 C.E.). Sustainable Development in Industry and Society 5.0: Governance, Management, and Financial Implications. In *https://services.igi-global.com/resolvedoi/resolve.aspx?doi=10.4018/979-8-3693-7322-4*. IGI Global. https://www.igi-global.com/book/sustainable-development-industry-society/www.igi-global.com/book/sustainable-development-industry-society/347660

Automobility in Transition?: A Socio-Technical Analysis of Sustainable Transport. (n.d.). Routledge & CRC Press. Retrieved 16 December 2024, from https://www.routledge.com/Automobility-in-Transition-A-Socio-Technical-Analysis-of-Sustainable-Transport/Geels-Kemp-Dudley-Lyons/p/book/9781032920382

Babkin, I., Alekseeva, N., Barabaner, H., & Antoshkova, N. (2022). *Forecasting Digital Economy Development Trends Based on Scientometric Data Monitoring*., DOI: 10.1007/978-3-030-81619-3_86

Barter, N., & Russell, S. (2012). *Sustainable Development: A Review Using Two Snapshots a Silver Jubilee Apart*. https://www.semanticscholar.org/paper/Sustainable-Development%3A-A-Review-Using-Two-a-Apart-Barter-Russell/c79392d9137a818d47dad99ae5b9629dd3dec463

Bril, A., Kalinina, O., & Valebnikova, O. (2016). *Innovation Venture Financing Projects in Information Technology* (Vol. 9870, p. 775). DOI: 10.1007/978-3-319-46301-8_67

Brouwer, A. S., van der Voet, E., Verkuijl, M. C., & Kramer, G. J. (2018). Journal of Cleaner Production. *Environmental Impacts of Electric Vehicle Battery Production and Recycling. RE:view, 200*, 915–934.

Cohen, B. D., & Kietzmann, J. (2014). *Ride on! Mobility Business Models for the Sharing Economy* (SSRN Scholarly Paper 2519349). Social Science Research Network. https://papers.ssrn.com/abstract=2519349 DOI: 10.1177/1086026614546199

Didenko, N., Skripnuk, D., Kikkas, K., Kalinina, O., & Kosinski, E. (2021). The Impact of Digital Transformation on the Micrologistic System, and the Open Innovation in Logistics. *Journal of Open Innovation, 7*(2), 115. DOI: 10.3390/joitmc7020115

Dubolazov, V., Tayushev, S., Gabdrakhmanova, I., Simakova, Z., & Leicht, O. (2022). Re-Engineering of Logistics Business Processes Influenced by the Digitalization. In Beskopylny, A., & Shamtsyan, M. (Eds.), *XIV International Scientific Conference "INTERAGROMASH 2021"* (Vol. 246, pp. 539–547). Springer International Publishing. DOI: 10.1007/978-3-030-81619-3_61

Fadeev, A., Petrov, I., Afanasiev, M., Melyokhin, V., & Livintsova, M. (2024). Formation of an Effective Transport and Logistics Strategy in the Mineral Fertilizers Industry in the Arctic. In K. S. Zokirjon ugli, A. Muratov, & S. Ignateva (Eds.), *Fundamental and Applied Scientific Research in the Development of Agriculture in the Far East (AFE-2022)* (pp. 175–185). Springer Nature Switzerland. DOI: 10.1007/978-3-031-37978-9_17

Huang, Y., Wang, H., Cao, G., Liu, J., & Liu, Y. (2020). Journal of Cleaner Production. *Environmental Impacts of Lithium-Ion Battery Production. RE:view, 257*, 120499.

Huntjens, P. (2021). *Towards a Natural Social Contract: Transformative Social-Ecological Innovation for a Sustainable, Healthy and Just Society*. Springer International Publishing., DOI: 10.1007/978-3-030-67130-3

Kalinina, O., Barykin, S., & Sergeev, M. NikolaevnaSemenova, G., Fatkullina, A., Mikhaylov, A., & De la Poza, E. (2022). Smart city perspectives in post-pandemic governance: Externalities reduction policy. *F1000Research, 11*, 1032. DOI: 10.12688/f1000research.123195.1

Lukashevich, N., Garanin, D., & Svirina, A. (2021, January 1). *(PDF) Multilevel prognosis of logistics chains in case of uncertainty: Information and statistical technologies implementation*. ResearchGate. https://www.researchgate.net/publication/322783975_Multilevel_prognosis_of_logistics_chains_in_case_of_uncertainty_Information_and_statistical_technologies_implementation

Management Association. I. R. (1 C.E.). Sustainable Development: Concepts, Methodologies, Tools, and Applications (3 Volumes). In *https://services.igi-global.com/resolvedoi/resolve.aspx?doi=10.4018/978-1-5225-3817-2*. IGI Global. https://www.igi-global.com/book/sustainable-development-concepts-methodologies-tools/www.igi-global.com/book/sustainable-development-concepts-methodologies-tools/181525

Popescu, C. R. G., Yu, P., & Wei, Y. (1 C.E.). Achieving the Sustainable Development Goals Through Infrastructure Development. In *https://services.igi-global.com/resolvedoi/resolve.aspx?doi=10.4018/979-8-3693-0794-6*. IGI Global. https://www.igi-global.com/book/achieving-sustainable-development-goals-through/www.igi-global.com/book/achieving-sustainable-development-goals-through/324723

Pupentsova, S., Demin, A., Kirilyuk, A., & Pupentsova, V. (2022). The Introduction of Digital Technologies to Participatory Design in the Public Spaces Formation. In Rodionov, D., Kudryavtseva, T., Skhvediani, A., & Berawi, M. A. (Eds.), *Innovations in Digital Economy* (pp. 325–342). Springer International Publishing., DOI: 10.1007/978-3-031-14985-6_23

Vashkevich, N., Barykin, S. E., Krasnov, A., Kurochkina, A. A., Alekseeva, N., Krasnova, D. S., Kankhva, V. S., Khovalova, T. V., Mikhov, O., Buniak, V., Dzhamaludinova, M. Y., Dedyukhina, N., & Ali, M. (2024). E-government based on sustainable digitalization: The case of the Chinese social sustainability marketing technologies paradigm. *Journal of Infrastructure. Policy and Development, 8*(3), 3. Advance online publication. DOI: 10.24294/jipd.v8i3.3113

Verbong, G. P. J., & Geels, F. W. (2010). Exploring sustainability transitions in the electricity sector with socio-technical pathways. *Technological Forecasting and Social Change, 77*(8), 1214–1221. DOI: 10.1016/j.techfore.2010.04.008

Vuković, D., Vlasova, N. Y., & Ergunova, O. T. (2020). Institutions in the Context of the Regional Development of Russia. In Gabdrakhmanov, N. K., & Safiullin, L. N. (Eds.), *Regional Economic Development in Russia* (pp. 123–132). Springer International Publishing., DOI: 10.1007/978-3-030-39859-0_11

Zhang, W., Barykin, S., Kirillova, T., Kapustina, I., Lukashevich, N., & Zaytsev, A. (2023). An Innovative Resource Management Framework using Logit-Boosted Machine Learning Algorithms for Vehicular Ad Hoc Networks (VANETs). *Engineering and Science*. Advance online publication. DOI: 10.30919/es980

Zhang, Y., Chen, Y., Li, X., Wang, Y., Wu, X., & Gao, Y. (2020). Journal of Cleaner Production. *A Review of Battery Recycling and Its Effect on Environment and Economy, 253*, 119886.

Compilation of References

Ab Rashid, M. F. (2023). How to conduct a bibliometric analysis using R packages: A comprehensive guidelines. *Journal of Tourism. Hospitality and Culinary Arts*, *15*(1), 24–39.

Abdelhakim, A., Soeiro, T. B., Stecca, M., & Canales, F. (2023). Multiport hybrid converter for electrified transportation systems. *IEEE Transactions on Industrial Electronics*, *70*(7), 6819–6829. DOI: 10.1109/TIE.2022.3199857

Adewale, T., & Paul, J. (2024). [and IoT: How These Technologies Are Creating the Perfect Storm for Smart Systems.]. *AI*, •••, 5G.

Afsar, B., Badir, Y., & Kiani, U. S. (2016). Linking spiritual leadership and employee pro-environmental behavior: The influence of workplace spirituality, intrinsic motivation, and environmental passion. *Journal of Environmental Psychology*, *45*, 79–88. DOI: 10.1016/j.jenvp.2015.11.011

Aftabuzzaman, M. (2007). Measuring traffic congestion: A critical review. In Proceedings of the 30th Australasian Transport Research Forum (pp. 1–16). Melbourne, Australia: Engineers Australia.

Aftabuzzaman, M., Currie, G., & Sarvi, M. (2010). Evaluating the congestion relief impacts of public transport in monetary terms. *Journal of Public Transportation*, *13*(1), 1–24. DOI: 10.5038/2375-0901.13.1.1

Aghabayk, K., Esmailpour, J., & Shiwakoti, N. (2021). Effects of COVID-19 on rail passengers' crowding perceptions. *Transportation Research Part A, Policy and Practice*, *154*, 186–202. https://doi.org/https://doi.org/10.1016/j.tra.2021.10.011. DOI: 10.1016/j.tra.2021.10.011 PMID: 34707330

Agrawal, S., Oza, P., Kakkar, R., Tanwar, S., Jetani, V., Undhad, J., & Singh, A. (2024). Analysis and recommendation system-based on PRISMA checklist to write systematic review. *Assessing Writing*, *61*, 100866. DOI: 10.1016/j.asw.2024.100866

Ahuja, R. K., Magnanti, T. L., & Orlin, J. B. (1993). *Network flows: Theory, algorithms, and applications*. Prentice Hall.

Aithal, P. S., & Aithal, S. (2022). Opportunities and Challenges for Green and Eco-Friendly Nanotechnology in Twenty-First Century. *Sustainable nanotechnology: Strategies, products, and applications*, 31-50. DOI: 10.1002/9781119650294.ch3

Alam, T., Gupta, R., Nasurudeen Ahamed, N., Ullah, A., & Almaghthwi, A. (2024). Smart mobility adoption in sustainable smart cities to establish a growing ecosystem: Challenges and opportunities. *MRS Energy & Sustainability : a Review Journal, 11*(2), 304–316. DOI: 10.1557/s43581-024-00092-4

Alanazi, F. (2023). Electric vehicles: Benefits, challenges, and potential solutions for widespread adaptation. *Applied Sciences (Basel, Switzerland), 13*(10), 6016. DOI: 10.3390/app13106016

Albino, V., Berardi, U., & Dangelico, R. M. (2015). Smart cities: Definitions, dimensions, performance, and initiatives. *Journal of Urban Technology, 22*(1), 3–21. DOI: 10.1080/10630732.2014.942092

Almatar, K. M. (2024). Increasing electric vehicles infrastructure in urban areas for efficiently employing renewable energy. *Environment, Development and Sustainability, 26*(10), 26183–26204. DOI: 10.1007/s10668-023-03723-x

Al-Rahamneh, A., Astrain, J. J., Villadangos, J., Klaina, H., Guembe, I. P., Lopez-Iturri, P., & Falcone, F. (2022). Bi2Bi communication: Toward encouragement of sustainable smart mobility. *IEEE Access : Practical Innovations, Open Solutions, 10*, 9380–9394. DOI: 10.1109/ACCESS.2022.3144643

Al-Thani, H., Koç, M., Isaifan, R. J., & Bicer, Y. (2022). A review of the integrated renewable energy systems for sustainable urban mobility. *Sustainability (Basel), 14*(17), 10517. DOI: 10.3390/su141710517

Amara, E., Moussa-Jerbi, A., & Tanfous, F. (2024). *Stakeholder Dynamics in the Context of a Local Urban Project*. Social Business., DOI: 10.1362/204440824X170747761359874

Anderson, M. L. (2014). Subways, strikes, and slowdowns: The impacts of public transit on traffic congestion. *The American Economic Review, 104*(9), 2763–2796. DOI: 10.1257/aer.104.9.2763

Anke, J., Francke, A., Schaefer, L., & Petzoldt, T. (2021). Impact of SARS-CoV-2 on the mobility behaviour in Germany. *European Transport Research Review, 13*(1), 10. Advance online publication. DOI: 10.1186/s12544-021-00469-3 PMID: 38624595

Anwar, S., Zia, M. Y. I., Rashid, M., de Rubens, G. Z., & Enevoldsen, P. (2020). Towards Ferry Electrification in the Maritime Sector. *Energies*, *13*(24), 6506. DOI: 10.3390/en13246506

Applegate, D. L., Bixby, R. E., Chvátal, V., & Cook, W. J. (2006). *The traveling salesman problem: A computational study*. Princeton University Press.

Arbabzadeh, M., & Kermani, M. J. (2019). Battery Recycling and Its Effect on Environment. In *Handbook of Materials Failure Analysis with Case Studies from the Aerospace and Automotive Industries* (pp. 561–575). Butterworth-Heinemann.
Atiku, S. O., Jeremiah, A., Semente, E., & Boateng, F. (1 C.E.). Sustainable Development in Industry and Society 5.0: Governance, Management, and Financial Implications. In *https://services.igi-global.com/resolvedoi/resolve.aspx?doi=10.4018/979-8-3693-7322-4*. IGI Global. https://www.igi-global.com/book/sustainable-development-industry-society/www.igi-global.com/book/sustainable-development-industry-society/347660

Arsenio, E., Martens, K., & Di Ciommo, F. (2016). Sustainable urban mobility plans: Bridging climate change and equity targets? *Research in Transportation Economics*, *55*, 30–39. DOI: 10.1016/j.retrec.2016.04.008

Arute, F., Arya, K., Babbush, R., Bacon, D., Bardin, J. C., Barends, R., Biswas, R., Boixo, S., Brandao, F. G. S. L., Buell, D. A., Burkett, B., Chen, Y., Chen, Z., Chiaro, B., Collins, R., Courtney, W., Dunsworth, A., Farhi, E., Foxen, B., & Martinis, J. M. (2019). Quantum supremacy using a programmable superconducting processor. *Nature*, *574*(7779), 505–510. DOI: 10.1038/s41586-019-1666-5 PMID: 31645734

Asadi, M., Fathy, M., Mahini, H., & Rahmani, A. M. (2022). An evolutionary game approach to safety-aware speed recommendation in fog/cloud-based intelligent transportation systems. *IEEE Transactions on Intelligent Transportation Systems*, *23*(7), 7431–7440. DOI: 10.1109/TITS.2021.3069850

Automobility in Transition?: A Socio-Technical Analysis of Sustainable Transport. (n.d.). Routledge & CRC Press. Retrieved 16 December 2024, from https://www.routledge.com/Automobility-in-Transition-A-Socio-Technical-Analysis-of-Sustainable-Transport/Geels-Kemp-Dudley-Lyons/p/book/9781032920382

Babkin, I., Alekseeva, N., Barabaner, H., & Antoshkova, N. (2022). *Forecasting Digital Economy Development Trends Based on Scientometric Data Monitoring*., DOI: 10.1007/978-3-030-81619-3_86

Bahl, A., & Dietzen, A. (2020). *Work-based learning as a driver for skills development and job creation*. International Labour Organization.

Banister, D. (2019). *Transport for sustainable development*. Routledge.

Bansal, P., & Graham, D. J. (2023). Congestion in cities: Can road capacity expansions provide a solution? *Transportation Research Part A, Policy and Practice, 174*, 103726. DOI: 10.1016/j.tra.2023.103726

Barman, P., Dutta, L., Bordoloi, S., Kalita, A., Buragohain, P., Bharali, S., & Azzopardi, B. (2023). Renewable energy integration with electric vehicle technology: A review of the existing smart charging approaches. *Renewable & Sustainable Energy Reviews, 183*, 113518. DOI: 10.1016/j.rser.2023.113518

Barter, N., & Russell, S. (2012). *Sustainable Development: A Review Using Two Snapshots a Silver Jubilee Apart*. https://www.semanticscholar.org/paper/Sustainable-Development%3A-A-Review-Using-Two-a-Apart-Barter-Russell/c79392d9137a818d47dad99ae5b9629dd3dec463

Batarseh, F. A., & Freeman, L. A. (2021). *AI and machine learning for transport planning and operations*. Elsevier.

Batty, M., Axhausen, K. W., Giannotti, F., Pozdnoukhov, A., Bazzani, A., Wachowicz, M., Ouzounis, G., & Portugali, Y. (2012). Smart cities of the future. *The European Physical Journal. Special Topics, 214*(1), 481–518. DOI: 10.1140/epjst/e2012-01703-3

Behruz, H., Chavoshy, A. P., Lavasani rad, A., & Mozaffari, G. (2013). Challenges of implementation of intelligent transportation systems in developing countries: Case study - Tehran. WIT Transactions on Ecology and the Environment, 179 VOLUME 2, 977–987. https://doi.org/DOI: 10.2495/SC130832

Bell, M. G. H., & Iida, Y. (1997). *Transportation network analysis*. John Wiley & Sons. DOI: 10.1002/9781118903032

Bertsimas, D., & Tsitsiklis, J. (2020). *Introduction to linear optimization*. Athena Scientific.

Bi, H., Ye, Z., & Zhu, H. (2022). Data-driven analysis of weather impacts on urban traffic conditions at the city level. *Urban Climate, 41*, 101065. DOI: 10.1016/j.uclim.2021.101065

Birch, D., Memery, J., & Kanakaratne, M. D. S. (2018). The mindful consumer: Balancing egoistic and altruistic motivations to purchase local food. *Journal of Retailing and Consumer Services, 40*, 221–228. DOI: 10.1016/j.jretconser.2017.10.013

Board, T. R. of Sciences Engineering, & Medicine. (2013). Deployment, Use, and Effect of Real-Time Traveler Information Systems (E. Robinson, T. Jacobs, K. Frankle, N. Serulle, & M. Pack, Eds.). The National Academies Press. https://doi.org/DOI: 10.17226/22664

Boccaletti, S., Bianconi, G., Criado, R., del Genio, C. I., Gómez-Gardeñes, J., Romance, M., Sendiña-Nadal, I., Wang, Z., & Zanin, M. (2014). The structure and dynamics of multilayer networks. *Physics Reports*, *544*(1), 1–122. DOI: 10.1016/j.physrep.2014.07.001 PMID: 32834429

Bohnsack, R., & Pinkse, J. (2017). Value propositions for disruptive technologies: Reconfiguration tactics in the case of electric vehicles. *California Management Review*, *59*(4), 79–96. DOI: 10.1177/0008125617717711

Bondemark, A., Andersson, H., Wretstrand, A., & Brundell-Freij, K. (2021). Is it expensive to be poor? Public transport in Sweden. *Transportation*, *48*(5), 2709–2734. DOI: 10.1007/s11116-020-10145-5

Borchers, T., Wittowsky, D., & Souza Fernandes, R. A. (2024). A comprehensive survey and future directions on optimising sustainable urban mobility. *IEEE Access : Practical Innovations, Open Solutions*, *12*, 63023–63048. DOI: 10.1109/ACCESS.2024.3393470

Borgosano, S., Martini, D., Longo, M., & Foiadelli, F. (2024). Electrifying urban transportation: A comparative study of battery swap stations and charging infrastructure for taxis in Chicago. *IEEE Access : Practical Innovations, Open Solutions*, *12*, 48017–48026. DOI: 10.1109/ACCESS.2024.3379138

Brake. (n.d.). CONNECTED AND AUTONOMOUS VEHICLES. Retrieved from https://www.brake.org.uk/get-involved/take-action/mybrake/knowledge-centre/vehicles/connected-and-autonomous-vehicles, Nov 20, 2024.

Bril, A., Kalinina, O., & Valebnikova, O. (2016). *Innovation Venture Financing Projects in Information Technology* (Vol. 9870, p. 775). DOI: 10.1007/978-3-319-46301-8_67

Broom, D. (2021). 4 reasons why electric cars haven't taken off yet. Retrieved from https://www.weforum.org/agenda/2021/07/electric-cars-batteries-fossil-fuel, 08 July.

Brouwer, A. S., van der Voet, E., Verkuijl, M. C., & Kramer, G. J. (2018). Journal of Cleaner Production. *Environmental Impacts of Electric Vehicle Battery Production and Recycling. RE:view*, *200*, 915–934.

Brown, J., Smith, L., & White, M. (2023). Drone technology for last-mile delivery: Opportunities and challenges. *Transportation Research Journal*, *12*(3), 45–67.

Brzeziński, Ł. (2024). Social Aspects of Smart Urban Mobility. *Encyclopedia.*, *4*(2), 864–873. DOI: 10.3390/encyclopedia4020055

Budiman, A. C., Fajratama, M. D., Ghaisani, D. N., Nurdini, A., Nurcahyo, R., & Habiburrahman, M. (2024, January). A study on electric motorcycle riding experiences and satisfaction: Perspectives from ride-hailing drivers for green urban mobility. In *2024 ASU International Conference in Emerging Technologies for Sustainability and Intelligent Systems (ICETSIS)* (pp. 1188–1191). DOI: 10.1109/ICETSIS61505.2024.10459358

Bukar, U. A., Sayeed, M. S., Razak, S. F. A., Yogarayan, S., Amodu, O. A., & Mahmood, R. A. R. (2023). A method for analyzing text using VOSviewer. *MethodsX*, *11*, 102339. DOI: 10.1016/j.mex.2023.102339 PMID: 37693657

Bureau of Labor Statistics. (2023). *Occupational outlook for heavy and tractor-trailer truck drivers.*

Butler, L., Yigitcanlar, T., & Paz, A. (2020). Smart urban mobility innovations: A comprehensive review and evaluation. *IEEE Access : Practical Innovations, Open Solutions*, *8*, 196034–196049. DOI: 10.1109/ACCESS.2020.3034596

Bylinko, L. (2024). Integration of an urban transportation system as an element of mobility plans. *Transport Problems*, *19*(1), 5–17. DOI: 10.20858/tp.2023.19.1.01

Cats, O., & Jenelius, E. (2018). Beyond a complete failure: The impact of partial capacity degradation on public transport network vulnerability. *Transport Policy*, *72*, 102–109.

Cats, O., Susilo, Y. O., & Reimal, T. (2017). The prospects of fare-free public transport: Evidence from Tallinn. *Transportation*, *44*(5), 1083–1104. DOI: 10.1007/s11116-016-9695-5

CB Insights. (2022). *Autonomous trucks: 40+ companies working on driverless trucking.*

Ceder, A. (2021). Urban mobility and public transport: Future perspectives and review. *International Journal of Urban Sciences*, *25*(4), 455–479. DOI: 10.1080/12265934.2020.1799846

Challenger, R., & Clegg, C. W. (2013). Crowd disasters: a socio-technical systems perspective. ln: Drury, J., Stott, C.J.T. (Eds.), Crowds in the 21st century. Perspectives from contemporary social science. Routledge, Ablngdon, Oxfordshire OX14 4RN, United Kingdom, pp. 80-97

Chambers, J., Barton, J., & Elser, M. (2020). *The circular economy and green jobs in transport*. International Transport Forum.

Chawla, U., Mohnot, R., Mishra, V., Singh, H. V., & Singh, A. K. (2023). Factors influencing customer preference and adoption of electric vehicles in India: A journey towards more sustainable transportation. *Sustainability (Basel), 15*(8), 7020. DOI: 10.3390/su15087020

Chen, C., Lan, Z., Zhan, G., Lyu, Y., Nie, B., & Li, S. E. (2024). Quantifying the individual differences of drivers' risk perception via potential damage risk model. *IEEE Transactions on Intelligent Transportation Systems, 25*(7), 8093–8104. DOI: 10.1109/TITS.2024.3379573

Chen, H., Li, X., & Zhao, Y. (2022). Artificial intelligence in transportation: Applications and future perspectives. *International Journal of Smart Transportation, 8*(2), 112–130.

Chen, L., Xu, X., & Zhang, Y. (2021). AI-driven transportation systems: Challenges and opportunities. *IEEE Transactions on Intelligent Transportation Systems, 22*(4), 1872–1884.

Chen, X., Qi, Y., & Liu, G. (2013). Empirical Study of Gap-Acceptance Behavior of Right-Turn-on-Red Drivers on Dual Right-Turn Lanes. *Journal of Transportation Engineering, 139*(2), 173–180. DOI: 10.1061/(ASCE)TE.1943-5436.0000489

Climate Change Commission. (n.d.) National Climate Change Action Plan 2011 – 2028. Retrieved from ((https://climate.emb.gov.ph/wp-content/uploads/2016/06/NCCAP-1.pdf

Coffman, M., Bernstein, P., & Wee, S. (2017). Electric vehicles revisited: A review of factors that affect adoption. *Transport Reviews, 37*(1), 79–93. DOI: 10.1080/01441647.2016.1217282

Cohen, B. D., & Kietzmann, J. (2014). *Ride on! Mobility Business Models for the Sharing Economy* (SSRN Scholarly Paper 2519349). Social Science Research Network. https://papers.ssrn.com/abstract=2519349 DOI: 10.1177/1086026614546199

Çolak, S., Lima, A., & González, M. C. (2016). Understanding congested travel in urban areas. *Nature Communications, 7*(1), 10793. DOI: 10.1038/ncomms10793 PMID: 26978719

Coletta, C., Heaphy, L., & Kitchin, R. (2019). From the accidental to articulated smart city: The creation and work of 'Smart Dublin'. *European Urban and Regional Studies, 26*(4), 349–364. DOI: 10.1177/0969776418785214

Colonel, M. (2020). Electrification of transportation sector nears tipping point. retrieved from https://www.renewableenergyworld.com/opinion-and-commentary/electrification-of-transportation-sector-nears-tipping-point/#gref, 01 July.

Cormen, T. H., Leiserson, C. E., Rivest, R. L., & Stein, C. (2009). *Introduction to algorithms*. MIT Press.

Correa, V. C., Lugo-Agudelo, L. H., Aguirre-Acevedo, D. C., Contreras, J. A. P., Borrero, A. M. P., Patiño-Lugo, D. F., & Valencia, D. A. C. (2020). Individual, health system, and contextual barriers and facilitators for the implementation of clinical practice guidelines: A systematic metareview. *Health Research Policy and Systems*, *18*(1), 74. DOI: 10.1186/s12961-020-00588-8 PMID: 32600417

Corti, F., Iacono, S. D., Astolfi, D., Pasetti, M., Vasile, A., Reatti, A., & Flammini, A. (2024). A comprehensive review of charging infrastructure for Electric Micro-mobility Vehicles: Technologies and challenges. *Energy Reports*, *12*, 545–567. DOI: 10.1016/j.egyr.2024.06.026

Cugurullo, F. (2020). Urban artificial intelligence: From automation to autonomy in the smart city. *Frontiers in Sustainable Cities*, *2*, 38. DOI: 10.3389/frsc.2020.00038

Daganzo, C. F. (2019). *Public transportation systems: Principles of system design, operations planning, and real-time control*. Springer. DOI: 10.1142/10553

Das, H. S., Rahman, M. M., Li, S., & Tan, C. W. (2020). Electric vehicles standards, charging infrastructure, and impact on grid integration: A technological review. *Renewable & Sustainable Energy Reviews*, *120*, 109618. DOI: 10.1016/j.rser.2019.109618

Davis, M. C., Challenger, R., Jayewardene, D. N. W., & Clegg, C. W. (2014). Advancing sociotechnical systems thinking: A call for bravery. *Applied Ergonomics*, *45*(2), 171–18. DOI: 10.1016/j.apergo.2013.02.009 PMID: 23664481

De Stefano, V. (2016). The rise of the "just-in-time workforce": On-demand work, crowdwork, and labor protection in the "gig-economy.". *Comparative Labor Law & Policy Journal*, *37*(3), 471–504.

Dediu, H. 2019. The micromobility definition. URL: https: //micromobility.io/news/the-micromobility-definition. Accessed: 2025-03-21

Department of Energy (DOE). (2021). Vehicle Technologies Office.

Dibra, L., Tafaj, L., & Borde, A. (2019). Striving to Adapt to Climate Change: Lessons from Albania. Retrieved form https://napglobalnetwork.org/2019/10/striving-to-adapt-to-climate-change-lessons-from-albania/, 29 October.

Didenko, N., Skripnuk, D., Kikkas, K., Kalinina, O., & Kosinski, E. (2021). The Impact of Digital Transformation on the Micrologistic System, and the Open Innovation in Logistics. *Journal of Open Innovation, 7*(2), 115. DOI: 10.3390/joitmc7020115

Dijkstra, E. W. (1959). A note on two problems in connexion with graphs. *Numerische Mathematik, 1*(1), 269–271. DOI: 10.1007/BF01386390

Dubolazov, V., Tayushev, S., Gabdrakhmanova, I., Simakova, Z., & Leicht, O. (2022). Re-Engineering of Logistics Business Processes Influenced by the Digitalization. In Beskopylny, A., & Shamtsyan, M. (Eds.), *XIV International Scientific Conference "INTERAGROMASH 2021"* (Vol. 246, pp. 539–547). Springer International Publishing. DOI: 10.1007/978-3-030-81619-3_61

Easymile. (2022) All Weather Performance for Autonomous Driving On The Road In Norway. Retrieved from https://easymile.com/news/all-weather-performance-autonomous-driving-road-norway, 01 March.

EgyptInnovate. (2021). Brightskies Reveal the 1st Autonomous Driving System in Egypt. Retrieved from https://egyptinnovate.com/en/news/brightskies-reveal-1st-autonomous-driving-system-egypt

Elassy, M., Al-Hattab, M., Takruri, M., & Badawi, S. (2024). Intelligent transportation systems for sustainable smart cities. *Transportation Engineering, 100252*, 100252. Advance online publication. DOI: 10.1016/j.treng.2024.100252

Electrive. (2021). US plans to drive transport electrification becoming concrete. Retrieved from https://www.electrive.com/2021/12/09/us-plans-to-drive-transport-electrification-becoming-concrete/#:~:text=The%20Biden%20administration%20is%20driving,be%20emission%2Dfree%20by%202027, 9 December.

Euler, L. (1736). *Solutio problematis ad geometriam situs pertinentis*. Commentarii Academiae Scientiarum Imperialis Petropolitanae.

European Commission. (2021). *Fit for 55: Delivering the EU's 2030 climate target on the way to climate neutrality.*

European Union. (2019). Artificial intelligence in transport: Current and future developments, opportunities, and challenges. Retrieved from https://www.europarl.europa.eu/RegData/etudes/BRIE/2019/635609/EPRS_BRI(2019)635609_EN.pdf

Fadeev, A., Petrov, I., Afanasiev, M., Melyokhin, V., & Livintsova, M. (2024). Formation of an Effective Transport and Logistics Strategy in the Mineral Fertilizers Industry in the Arctic. In K. S. Zokirjon ugli, A. Muratov, & S. Ignateva (Eds.), *Fundamental and Applied Scientific Research in the Development of Agriculture in the Far East (AFE-2022)* (pp. 175–185). Springer Nature Switzerland. DOI: 10.1007/978-3-031-37978-9_17

Falcocchio, J. C., & Levinson, H. S. (2015). *Road traffic congestion: a concise guide* (Vol. 7). Springer International Publishing. DOI: 10.1007/978-3-319-15165-6

Fan, L., Wang, J., Chang, Y., Li, Y., Wang, Y., & Cao, D. (2024). 4D mmWave radar for autonomous driving perception: A comprehensive survey. *IEEE Transactions on Intelligent Vehicles*, *9*(4), 4606–4620. DOI: 10.1109/TIV.2024.3380244

Fan, Y., Khattak, A. J., & Shay, E. (2007). Intelligent Transportation Systems: What Do Publications and Patents Tell Us? *Journal of Intelligent Transport Systems*, *11*(2), 91–103. DOI: 10.1080/15472450701301081

Fatima, H., & Rasool, M. (2025). AI-Driven Urban Mobility Solutions: Enhancing Smart Traffic Management and Urban Infrastructure Optimization. DOI: 10.13140/RG.2.2.25367.48803

Fielbaum, A., Ruiz, F., Boccardo, G., Rubio, D., Tirachini, A., & Ro, J. (2023). The job of public transport, ride-hailing and delivery drivers: Conditions during the COVID-19 pandemic and implications for a post-pandemic future. *Travel Behaviour & Society*, *31*, 63–77. DOI: 10.1016/j.tbs.2022.11.004 PMID: 36405769

Figenbaum, E. (2017). Perspectives on Norway's supercharged electric vehicle policy. *Environmental Innovation and Societal Transitions*, *25*, 14–34. DOI: 10.1016/j.eist.2016.11.002

Fjelstul, J. (25 July 2014). Vehicle electrification: On the "green" road to destination sustainability, *Journal of Destination Marketing & Management, 3*(3), 2014, Pages 137-139, ISSN 2212-571X, . (https://www.sciencedirect.com/science/article/pii/S2212571X14000316)DOI: 10.1016/j.jdmm.2014.06.002

Ford, L. R., & Fulkerson, D. R. (1962). *Flows in networks*. Princeton University Press.

Frangoul, A. (2019). Artificial intelligence is going to control on-demand bus services in Japan. Retrieved from https://www.cnbc.com/2019/03/13/artificial-intelligence-to-control-on-demand-bus-services-in-japan.html#:~:text=The%20Mitsubishi%20Corporation%20has%20set,a%20major%20Japanese%20bus%20operator, 13 March.

Frey, C. B., & Osborne, M. A. (2017). The future of employment: How susceptible are jobs to computerisation? *Technological Forecasting and Social Change*, *114*, 254–280. DOI: 10.1016/j.techfore.2016.08.019

Frikha, R., & Hlali, A. (2023). The Evolution Of Logistics Management Toward The Digital Transition. International Journal of Commerce. *Industry and Entrepreneurship Studies*, *3*, 66–75.

Furtado, F., Blanco, J. P., Pritchard, J. P., & Vazano, P. (n.d.). Policy Scenarios for Decarbonising Argentina's Transport System. retrieved from https://www.itf-oecd.org/sites/default/files/docs/policy-scenarios-decarbonising-transport-argentina.pdf

Garcia, P., & Lopez, R. (2021). The role of electric and hydrogen vehicles in sustainable transportation. *Sustainable Transport Review*, *9*(4), 201–215.

Ghaplanyan, I., & Grigoryan, E. (11 March 2021). The Prospects of Electric Mobility in Armenia. Retrieved from https://evnreport.com/economy/the-prospects-of-electric-mobility-in-armenia-2/

Gheorghe, C., & Soica, A. (2025). Revolutionizing Urban Mobility: A Systematic Review of AI, IoT, and Predictive Analytics in Adaptive Traffic Control Systems for Road Networks. *Electronics (Basel)*, *14*(4), 719. DOI: 10.3390/electronics14040719

Ghosh, A., Bose, S., & Mukherjee, P. (2021). Big data analytics in multi-modal transportation networks. *IEEE Transactions on Intelligent Transportation Systems*, *22*(7), 5678–5690.

Ghosh, D., & Choi, J. P. (2022). Cybersecurity challenges in smart transportation systems. *Transportation Research Part C, Emerging Technologies*, *140*, 103719.

Giannini, F., Franzè, G., Pupo, F., & Fortino, G. (2023). A sustainable multi-agent routing algorithm for vehicle platoons in urban networks. *IEEE Transactions on Intelligent Transportation Systems*, *24*(12), 14830–14840. DOI: 10.1109/TITS.2023.3305463

Gita-Carlos, R. A. (2022). Bills regulating vintage, electric vehicles lapse into law. Retrieved from https://www.pna.gov.ph/articles/1173013, 26 April.

Gittleman, M., & Monaco, K. (2020). *Truck automation and job displacement in the United States*. University of Pennsylvania.

Gkiotsalitis, K., & Cats, O. (2021). Public transport optimization using big data and machine learning techniques. *Transport Reviews*, *41*(3), 323–345.

Gkiotsalitis, K., & Cats, O. (2021). Public transport planning adaption under the COVID-19 pandemic crisis: Literature review of research needs and directions. *Transport Reviews*, *41*(3), 374–392. DOI: 10.1080/01441647.2020.1857886

Glaister, S., & Graham, D. (2003). Transport pricing and investment in England. Imperial college (Department of civil Engineering).

Gönül, Ö., Duman, A. C., & Güler, Ö. (2021). Electric vehicles and charging infrastructure in Turkey: An overview. *Renewable & Sustainable Energy Reviews*, *143*, 110913. DOI: 10.1016/j.rser.2021.110913

GOV.PH. (n.d.). Securing The Future of Philippine Industries. Retrieved from https://industry.gov.ph/industry/e-vehicles/

Graf, A., & Kantel, A. (2025). Planning sustainable and inclusive transformation: An energy democracy perspective on urban mobility. *Frontiers in Political Science*, *7*, 1421569. Advance online publication. DOI: 10.3389/fpos.2025.1421569

Graham, J. D., & Brungard, E. (2021). Consumer adoption of plug-in electric vehicles in selected countries. *Future Transportation*, *1*(2), 303–325. DOI: 10.3390/futuretransp1020018

Griffiths, S., Del Rio, D. F., & Sovacool, B. (2021). Policy mixes to achieve sustainable mobility after the COVID-19 crisis. *Renewable & Sustainable Energy Reviews*, *143*, 110919. DOI: 10.1016/j.rser.2021.110919 PMID: 35702384

Guimerà, R., Mossa, S., Turtschi, A., & Amaral, L. A. N. (2005). The worldwide air transportation network: Anomalous centrality, community structure, and cities' global roles. *Proceedings of the National Academy of Sciences of the United States of America*, *102*(22), 7794–7799. DOI: 10.1073/pnas.0407994102 PMID: 15911778

Gulc, A., & Budna, K. (2024). Classification of Smart and Sustainable Urban Mobility. *Energies*, *17*(9), 2148. DOI: 10.3390/en17092148

Gulzari, A., Wang, Y., & Prybutok, V. (2022). A green experience with eco-friendly cars: A young consumer electric vehicle rental behavioral model. *Journal of Retailing and Consumer Services*, *65*, 102877. DOI: 10.1016/j.jretconser.2021.102877

Guo, Z., Yu, K., Konstantin, K., Mumtaz, S., Wei, W., Shi, P., & Rodrigues, J. J. P. C. (2023). Deep collaborative intelligence-driven traffic forecasting in green internet of vehicles. *IEEE Transactions on Green Communications and Networking*, *7*(2), 1023–1035. DOI: 10.1109/TGCN.2022.3193849

Gupta, K., & Patel, S. (2021). IoT in transportation: Challenges and opportunities. *Journal of Emerging Technologies in Mobility*, *7*(3), 55–72.

Hansen, M., Huang, Y., & Wang, X. (2022). Multi-modal transport networks: A review of optimization techniques. *Transportation Research Part B: Methodological, 156*, 76–98.

Hardman, S., Chandan, A., Tal, G., & Turrentine, T. (2017). The effectiveness of financial purchase incentives for battery electric vehicles–A review of the evidence. *Renewable & Sustainable Energy Reviews, 80*, 1100–1111. DOI: 10.1016/j.rser.2017.05.255

Heumann, M., Kraschewski, T., Otto, P., Tilch, L., Brauner, T., & Breitner, M. (2025). Factors influencing the usage of shared micromobility: Implications from Berlin. *Journal of Cycling and Micromobility Research., 100063*, 100063. Advance online publication. DOI: 10.1016/j.jcmr.2025.100063

Hilmola, O. P., & Henttu, V. (2022). Transformation of logistics industry towards environmental sustainability and automation. *Sustainability, 14*(1), 118.

Hlali, A., & Gafsi, N. (2024). Analysis of Digitalization and Sustainable Development in Africa. *Perspectives on Global Development and Technology, 22*(5-6), 415–428. DOI: 10.1163/15691497-12341668

Hockenos, P. (2018). Europe Takes First Steps in Electrifying World's Shipping Fleets. Retrieved from https://e360.yale.edu/features/europe-takes-first-steps-in-electrifying-worlds-shipping-fleet, 22 February.

Hordnes, E. (2020). RACE TO ELECTRIFICATION – NORWAY IN A POLE POSITION. Retrieved from https://www.swecourbaninsight.com/urban-energy/race-to-electrification-norway-in-a-pole-position/#:~:text=By%202020%2C%20Norway%20will%20also,global%20battery%2Dpowered%20ship%20fleet, 31 August.

Hou, B., Bose, B., & Haran, K. (2020). Powering Electric Aircraft at O'Hare Airport: A Case Study. . Retrieved from https://ieeexplore.ieee.org/document/9281871, 16 December.DOI: 10.1109/PESGM41954.2020.9281871

Hou, B., Bose, S., Mela, L., & Haran, K. (2021). Impact of Aviation Electrification on Airports: Flight Scheduling and Charging. retrieved from https://arxiv.org/pdf/2108.08963.pdf, 20 August.

Huang, Y., Qian, L., Soopramanien, D., & Tyfield, D. (2021). Buy, lease, or share? Consumer preferences for innovative business models in the market for electric vehicles. *Technological Forecasting and Social Change, 166*, 120639. DOI: 10.1016/j.techfore.2021.120639

Huang, Y., Wang, H., Cao, G., Liu, J., & Liu, Y. (2020). Journal of Cleaner Production. *Environmental Impacts of Lithium-Ion Battery Production. RE:view, 257*, 120499.

Hui, A. (2019). Understanding the positioning of “the electric vehicle consumer”: Variations in interdisciplinary discourses and their implications for sustainable mobility systems. *Applied Mobilities*, *4*(1), 49–65. DOI: 10.1080/23800127.2017.1380977

Huntjens, P. (2021). *Towards a Natural Social Contract: Transformative Social-Ecological Innovation for a Sustainable, Healthy and Just Society*. Springer International Publishing., DOI: 10.1007/978-3-030-67130-3

Hussain, R., Son, J., Eun, H., & Oh, H. (2020). Internet of Vehicles: Challenges and future technologies. *IEEE Access : Practical Innovations, Open Solutions*, *8*, 146173–146197.

Hwang, S., Lee, K., Jeon, H., & Kum, D. (2022). Autonomous vehicle cut-in algorithm for lane-merging scenarios via policy-based reinforcement learning nested within finite-state machine. *IEEE Transactions on Intelligent Transportation Systems*, *23*(10), 17594–17606. DOI: 10.1109/TITS.2022.3153848

IEA. (2022). Renewables and energy efficiency can reduce Armenia’s dependence on energy imports, IEA policy review says. Retrieved from https://www.iea.org/news/renewables-and-energy-efficiency-can-reduce-armenia-s-dependence-on-energy-imports-iea-policy-review-says, 22 March.

IEA. (2023). *Global EV Outlook 2023: Catching up with climate ambitions*. International Energy Agency.

ILO. (2021). *World employment and social outlook: The role of digital labour platforms in transforming the world of work*. International Labour Organization.

International Energy Agency (IEA). (2022). *Global EV Outlook 2022*.

International Energy Agency (IEA). (2022).. . *Transport*.

International Energy Agency (IEA). (2023). *Global EV Outlook 2023*.

International Labour Organization (ILO). (2018). [*Greening with jobs*.]. *World Employment and Social Outlook*, •••, 2018.

International Transport Forum (ITF). (2017). *Managing the Transition to Driverless Road Freight Transport*.

Irvin, R. A., & Stansbury, J. (2004). Citizen participation in decision making: Is it worth the effort? *Public Administration Review*, *64*(1), 55–65. DOI: 10.1111/j.1540-6210.2004.00346.x

Jarayaman, K. (2019). Third and crucial era of aviation. Retrieved from https://www.dnaindia.com/analysis/column-third-and-crucial-era-of-aviation-2795694, 09 October.

Jarrar, O. M., & Al-Homoud, M. (2024). Sustainable urban development in Riyadh: A projected model for walkability. *International Journal of Urban Sustainable Development, 16*(1), 398–421. DOI: 10.1080/19463138.2024.2421163

Jenn, A. (2019). Equity impacts of electric vehicle policies in the United States. *Transportation Research Part D, Transport and Environment, 67*, 1–8.

Jiao, N., & Evans, S. (2016). Business models for sustainability: The case of second-life electric vehicle batteries. *Procedia CIRP, 40*, 250–255. DOI: 10.1016/j.procir.2016.01.114

Jing, Q.-L., Liu, H.-Z., Yu, W.-Q., & He, X. (2022). The Impact of Public Transportation on Carbon Emissions—From the Perspective of Energy Consumption. *Sustainability (Basel), 14*(10), 6248. DOI: 10.3390/su14106248

Jobin, A., Ienca, M., & Vayena, E. (2019). The global landscape of AI ethics guidelines. *Nature Machine Intelligence, 1*(9), 389–399. DOI: 10.1038/s42256-019-0088-2

Johnson, T., & White, C. (2023). Risk management strategies in transportation networks using big data. *Safety & Security in Transport, 11*(1), 98–115.

Joshi, N. (2019). How AI Can Transform The Transportation Industry. Retrieved from https://www.forbes.com/sites/cognitiveworld/2019/07/26/how-ai-can-transform-the-transportation-industry/?sh=1a2bf0b84964, 26 July.

Kalinina, O., Barykin, S., & Sergeev, M. NikolaevnaSemenova, G., Fatkullina, A., Mikhaylov, A., & De la Poza, E. (2022). Smart city perspectives in post-pandemic governance: Externalities reduction policy. *F1000Research, 11*, 1032. DOI: 10.12688/f1000research.123195.1

Kamargianni, M., Matyas, M., Li, W., & Muscat, J. (2016). *Londoners' attitudes towards car-ownership and Mobility-as-a-Service: Impact assessment and opportunities that lie ahead.* UCL Energy Institute.

Kang, J., Luo, X., Nie, J., Wu, T., Zhou, H., Wang, Y., Niyato, D., Mao, S., & Xie, S. (2024). Blockchain-based pseudonym management for vehicle twin migrations in vehicular edge metaverse. *IEEE Internet of Things Journal, 11*(21), 34254–34269. DOI: 10.1109/JIOT.2024.3404559

Kashem, M. A., Shamsuddoha, M., & Nasir, T. (2024). Sustainable Transportation Solutions for Intelligent Mobility: A Focus on Renewable Energy and Technological Advancements for Electric Vehicles (EVs) and Flying Cars. *Future Transportation, 4*(3), 874–890. DOI: 10.3390/futuretransp4030042

Khan, W., Ahmad, A., Ahmad, F., & Saad Alam, M. (2018). A comprehensive review of fast charging infrastructure for electric vehicles. *Smart Science, 6*(3), 256–270. DOI: 10.1080/23080477.2018.1437323

Kovačić, M., Mutavdžija, M., & Buntak, K. (2022). New paradigm of sustainable urban mobility: Electric and autonomous vehicles—A review and bibliometric analysis. *Sustainability (Basel), 14*(15), 9525. DOI: 10.3390/su14159525

Kumar, P., Kumar, R., Gupta, G. P., & Tripathi, R. (2022). BDEdge: Blockchain and deep-learning for secure edge-envisioned green CAVs. *IEEE Transactions on Green Communications and Networking, 6*(3), 1330–1339. DOI: 10.1109/TGCN.2022.3165692

Kumar, R. R., & Alok, K. (2020). Adoption of electric vehicle: A literature review and prospects for sustainability. *Journal of Cleaner Production, 253*, 119911. DOI: 10.1016/j.jclepro.2019.119911

Laporte, G., & Mesa, J. A. (1993). The design of rapid transit networks. *Location Science, 1*(2), 87–102.

Leavitt, H. J. (1965). Applied Organizational Change in Industry: Structural, Technological, and Humanistic Approaches. In March, J. G. (Ed.), *Handbook of Organizations, 1st* (pp. 1144–1170). Routledge.

Lee, S., & Kim, J. (2020). Blockchain applications in logistics and transportation. *Journal of Supply Chain Innovations, 6*(2), 133–150.

Leonchik, V. P., Dolinina, O. E., and Friedrich Ebert Stiftung (2019). Sustainable mobility in livable cities. Global challenges and new approaches: [handbook on transportation system planning]. Minsk: Kolograd, 32 p. (Analysis) (Climate change, energy and environment). Bibliography in footnotes. ISBN 978-985-596-473-6.

Li, H., Zhao, Y., Mao, Z., Qin, Y., Xiao, Z., Feng, J., Gu, Y., Ju, W., Luo, X., & Zhang, M. (2024). Graph Neural Networks in Intelligent Transportation Systems: Advances, Applications and Trends. arXiv preprint arXiv:2401.00713. https://arxiv.org/abs/2401.00713

Liang, X., Lee, L., Xing, E. P., & Xing, E. (2018). Graph-structured multi-modal fusion with structure-evolving LSTM. *Proceedings of the 24th ACM SIGKDD International Conference on Knowledge Discovery & Data Mining*, 1485–1494. https://arxiv.org/abs/1707.00750

Lin, W., Wei, H., Yang, L., & Zhao, X. (2024). Technical review of electric vehicle charging distribution models with considering driver behaviors impacts. [English Edition]. *Journal of Traffic and Transportation Engineering*, *11*(4), 643–666. https://doi.org/https://doi.org/10.1016/j.jtte.2024.06.001. DOI: 10.1016/j.jtte.2024.06.001

Liu, X., Sun, X., Zheng, H., & Huang, D. (2021). Do policy incentives drive electric vehicle adoption? Evidence from China. *Transportation Research Part A, Policy and Practice*, *150*, 49–62. DOI: 10.1016/j.tra.2021.05.013

Li, W., Wang, H., & Zhang, Y. (2020). IoT and transportation network integration: Challenges and solutions. *Journal of Transport and Mobility Research*, *8*(2), 112–130.

Li, Z., Liu, A., Shang, W.-L., Li, J., Lu, H., & Zhang, H. (2023). Sustainability assessment of regional transportation: An innovative fuzzy group decision-making model. *IEEE Transactions on Intelligent Transportation Systems*, *24*(12), 15959–15973. DOI: 10.1109/TITS.2023.3275141

Lohmann, G., & Duval, D. T. (2011). Critical Aspects of the Tourism-Transport Relationship. Goodfellow Publishers Ltd 2011. Retrieved from https://www.goodfellow publishers.com/free_files/fileTransportReview.pdf

López, I., Ibarra, E., Matallana, A., Andreu, J., & Kortabarria, I. (2019). Next generation electric drives for HEV/EV propulsion systems: Technology, trends and challenges. *Renewable & Sustainable Energy Reviews*, *114*, 109336. DOI: 10.1016/j.rser.2019.109336

Lou, Y., Sun, D., & Zhang, H. (2023). Graph theory applications in urban transportation: A review. *Journal of Advanced Transportation*, *2023*, 1–19.

Luca, O., Andrei, L., Iacoboaea, C., & Gaman, F. (2023). Unveiling the hidden effects of automated vehicles on "do no significant harm" components. *Sustainability (Basel)*, *15*(14), 11265. DOI: 10.3390/su151411265

Lucas, K. (2019). Transport and social exclusion: Where are we now? *Transport Policy*, *20*, 105–113. DOI: 10.1016/j.tranpol.2012.01.013

Lukashevich, N., Garanin, D., & Svirina, A. (2021, January 1). *(PDF) Multilevel prognosis of logistics chains in case of uncertainty: Information and statistical technologies implementation*. ResearchGate. https://www.researchgate.net/publication/322783975_Multilevel_prognosis_of_logistics_chains_in_case_of_uncertainty_Information_and_statistical_technologies_implementation

Luo, H., Chahine, R., Gkritza, K., & Cai, H. (2023). What motivates the use of shared mobility systems and their integration with public transit? Evidence from a choice experiment study. *Transportation Research Part C, Emerging Technologies*, *155*, 104286. DOI: 10.1016/j.trc.2023.104286

LYNX Strategic Business Advisors. (2020). EGYPT's EVOLVING POLICY FRAMEWORK FOR ELECTRIC VEHICLES. Retrieved from https://www.lynxegypt.com/assets/pdfs/EVs-Policy-Regulatory.pdf, November 20.

Ma, C. & Vu, H. & Wang, J. & Trieu, V.H. & Li, G. (2023). Understanding Residents' Behavior for Smart City Management by Sequential and Periodic Pattern Mining. IEEE Transactions on Computational Social Systems. PP. 1-17. DOI: 10.1109/TCSS.2023.3249740

MacEacheran, M. (2019). Start of aviation's third era? Zero-emission seaplanes may soon be a reality. Retrieved from https://www.business-standard.com/article/companies/start-of-aviation-s-third-era-zero-emission-seaplane-may-soon-be-a-reality-119111601300_1.html, 16 November.

Mammadov, R. (2012). The Importance of Transportation in Tourism Sector. 7th Silk Road International Conference "Challenges and Opportunities of Sustainable Economic Development in Eurasian Countries", May 24, 2012 – May 26, 2012. *SSRN*, *24*(May). https://ssrn.com/abstract=223133

Management Association. I. R. (1 C.E.). Sustainable Development: Concepts, Methodologies, Tools, and Applications (3 Volumes). In *https://services.igi-global.com/resolvedoi/resolve.aspx?doi=10.4018/978-1-5225-3817-2*. IGI Global. https://www.igi-global.com/book/sustainable-development-concepts-methodologies-tools/www.igi-global.com/book/sustainable-development-concepts-methodologies-tools/181525

Manda, V. K., Christy, V., & Hlali, A. (2025). Current trends, opportunities, and future research directions in geospatial technologies for smart cities. In Recent Trends in Geospatial AI (pp. 239–270).

Mandys, F. (2021). Electric vehicles and consumer choices. *Renewable & Sustainable Energy Reviews*, *142*, 110874. DOI: 10.1016/j.rser.2021.110874

Manyika, J., Lund, S., Chui, M., Bughin, J., Woetzel, J., Batra, P., & Sanghvi, S. (2017). *Jobs lost, jobs gained: Workforce transitions in a time of automation.* McKinsey Global Institute.

Marr, B. (2022). The 3 Biggest Future Trends In Transportation And Mobility. Retrieved from https://www.forbes.com/sites/bernardmarr/2022/01/20/the-3-biggest-future-trends-in-transportation-and-mobility/?sh=3d84087b3783, 20 January.

Martens, K. (2017). *Transport justice: Designing fair transportation systems.* Routledge.

Marve, S. R., Bhorkar, M., & Baitule, P. (2016). A survey on environmental impacts due to traffic congestion in peak hours. IJSTE, 2(08).

Mastoi, M. S., Zhuang, S., Munir, H. M., Haris, M., Hassan, M., Usman, M., Bukhari, S. S. H., & Ro, J. S. (2022). An in-depth analysis of electric vehicle charging station infrastructure, policy implications, and future trends. *Energy Reports*, *8*, 11504–11529. DOI: 10.1016/j.egyr.2022.09.011

Mateescu, C., & Popa, I. (2017). European Best Practices and Policies in Promoting Green Mobility. *EEA – Electrotehnica, Electronica. Automatica*, *65*(4), 12–16.

Maurya, S., Shavkidinova, D., Shrivastava, V., & Subrahmanyam, S. (2025). Effects of Green Supply Chain Management Practices on Environmental and Firm Sustainability Performance. *ACTA INNOVATIONS*, *55*, 13–22.

Mavlutova, I., Atstaja, D., Grasis, J., Kuzmina, J., Uvarova, I., & Roga, D. (2023). Urban transportation concept and sustainable urban mobility in smart cities: A review. *Energies*, *16*(8), 3585. DOI: 10.3390/en16083585

McKinsey & Company. (2022). *The future of work in transportation: Preparing for technological disruption.* https://www.mckinsey.com/industries/travel-logistics-and-infrastructure/our-insights/the-future-of-work-in-transportation

McKinsey & Company. (2023). *The future of automation in logistics: Robotics and drones.*

Metz, D. (2018). Tackling urban traffic congestion: The experience of London, Stockholm and Singapore. *Case Studies on Transport Policy*, *6*(4), 494–498. DOI: 10.1016/j.cstp.2018.06.002

Milakis, D., van Arem, B., & van Wee, B. (2017). Policy and society related implications of automated driving: A review of literature and directions for future research. *Journal of Intelligent Transport Systems*, *21*(4), 324–348. DOI: 10.1080/15472450.2017.1291351

Miles, J. C. (2014). Intelligent Transport Systems: Overview and Structure (History, Applications, and Architectures). In *Major Reference Works*. Wiley Online Library., DOI: 10.1002/9781118354179.auto166

Ministry for the Environment. (2022). Adapting to climate change. retrieved from https://environment.govt.nz/what-government-is-doing/areas-of-work/climate-change/adapting-to-climate-change/, 11 April.

Ministry of Climate and Environment. (2021). Norway's first adaptation communication. Retrieved from https://unfccc.int/sites/default/files/resource/Adaptation%20Communication%20Norway.pdf, 05 March.

Ministry of Manpower Singapore. (2021). *Strengthening protections for platform workers.*

Mishal, A., Dubey, R., Gupta, O. K., & Luo, Z. (2017). Dynamics of environmental consciousness and green purchase behaviour: An empirical study. *International Journal of Climate Change Strategies and Management*, *9*(5), 682–706. DOI: 10.1108/IJCCSM-11-2016-0168

Mishra, S., Rajendran, P. K., Vecchietti, L. F., & Har, D. (2023). Sensing accident-prone features in urban scenes for proactive driving and accident prevention. *IEEE Transactions on Intelligent Transportation Systems*, *24*(9), 9401–9414. DOI: 10.1109/TITS.2023.3271395

Miskolczi, M., Földes, D., Munkacsy, A., & Jaszberenyi, M. (2021). Urban mobility scenarios until the 2030s. *Sustainable Cities and Society*, *72*, 103029. DOI: 10.1016/j.scs.2021.103029

Mitici, M., Pereira, M., & Oliviero, F. (2022). Electric flight scheduling with battery-charging and battery-swapping opportunities, EURO Journal on Transportation and Logistics, Volume 11, 2022, 100074, ISSN 2192-4376, . (https://www.sciencedirect.com/science/article/pii/S2192437622000036), 16 February.DOI: 10.1016/j.ejtl.2022.100074

Mkrtchyan, N. (2024). Drawing of the Russian population into large cities and their suburbs. *Journal of the New Economic Association, New Economic Association*, *63*(2), 241–248. DOI: 10.31737/22212264_2024_2_241-248

Mofolasayo, A. (2023). Assessing and Managing the Direct and Indirect Emissions from Electric and Fossil-Powered Vehicles. *Sustainability (Basel)*, *15*(2), 1138. DOI: 10.3390/su15021138

Mohamed, M., Ferguson, M., & Ahmad, M. (2020). The role of artificial intelligence in enhancing public transport systems. *Transportation Research Interdisciplinary Perspectives, 4*, 100091.

Mohtich, F.-E., Bensiali, S., Idr, A., & El-Ayachi, M. (2023). Geospatial data integration for intelligent transportation systems: A review of the state of the art of spatio-temporal traffic forecasting. *African Journal on Land Policy and Geospatial Sciences, 6*(4), 683–692. DOI: 10.48346/IMIST.PRSM/ajlp-gs.v6i4.41748

Morton, C., Anable, J., & Nelson, J. D. (2017). Consumer structure in the emerging market for electric vehicles: Identifying market segments using cluster analysis. *International Journal of Sustainable Transportation, 11*(6), 443–459. DOI: 10.1080/15568318.2016.1266533

Moser, D. (2021). Approaches to transformative urban mobility. *Journal of Urban Regeneration and Renewal, 14*(3), 255. Advance online publication. DOI: 10.69554/CNPJ9514

Mukherjee, A., Singh, R., & Banerjee, D. (2021). Enhancing security in freight transport using blockchain. *Logistics & Supply Chain Management Journal, 10*(1), 44–60.

Müller-Eie, D., & Kosmidis, I. (2023). Sustainable mobility in smart cities: A document study of mobility initiatives of mid-sized Nordic smart cities. *European Transport Research Review, 15*(1), 36. DOI: 10.1186/s12544-023-00610-4

Mulley, C., & Nelson, J. D. (2020). Mobility as a Service (MaaS) – does it have critical mass? *Transport Reviews, 40*(3), 239–243.

Mykhailova, K. (2023). *Implementation of urban mobility projects for smart cities: approaches to justification and evaluation.* Black Sea Economic Studies., DOI: 10.32782/bses.84-20

Narayanan, S., & Antoniou, C. (2023). Shared mobility services towards Mobility as a Service (MaaS): What, who and when? *Transportation Research Part A, Policy and Practice, 168*, 103581. Advance online publication. DOI: 10.1016/j.tra.2023.103581

Newman, M. E. J. (2003). The structure and function of complex networks. *SIAM Review, 45*(2), 167–256. DOI: 10.1137/S003614450342480

Newman, M. E. J. (2010). *Networks: An Introduction.* Oxford University Press. DOI: 10.1093/acprof:oso/9780199206650.001.0001

Newman, M. E. J. (2018). *Networks: An introduction.* Oxford University Press. DOI: 10.1093/oso/9780198805090.001.0001

Nguyen, D. M., Kishk, M. A., & Alouini, M.-S. (2021). Modeling and analysis of dynamic charging for EVs: A stochastic geometry approach. *IEEE Open Journal of Vehicular Technology, 2*, 17–44. DOI: 10.1109/OJVT.2020.3032588

Ning, X., Tian, H., Lin, Y., Yao, X., Hu, F., & Yin, Y. (2024). Research on multi-objective optimization models for intersection crossing of connected autonomous vehicles with traffic signals. *IEEE Access : Practical Innovations, Open Solutions, 12*, 36825–36840. DOI: 10.1109/ACCESS.2024.3374041

Nurulin, Y., Skvortsova, I., & Kaltchenko, O. Charging Infrastructure for Electric Vehicles: Problems and Development Prospects. Lecture Notes in Networks and Systems, 2021, 460 LNNSP, pp. 239–248. DOI: 10.1007/978-3-031-20875-1_22

Nurulin, Y. R., Skvortsova, I., & Vinogradova, E. (2020). On the issue of the green energy markets development. *Lecture Notes in Networks and Systems, 95*, 360–367. DOI: 10.1007/978-3-030-34983-7_34

O'Neill-Carrillo, E.; Lave, M.; Haines, T. (2021). Systemwide Considerations for Electrification of Transportation in Islands and Remote Locations. *Vehicles*. 3, 498–511., 6 August.DOI: 10.3390/vehicles3030030

OECD. (2019). *Future of Education and Skills 2030: Curriculum analysis*. Organisation for Economic Co-operation and Development.

OECD. (2020). *The digital transformation of SMEs*.

OECD. (2022). *Skills for a 21st century transport workforce*. Organisation for Economic Co-operation and Development.

OECD. (2022). *Skills for the digital transformation of logistics and transport*.

Okoh, A. S., & Onuoha, M. C. (2024). Immediate and future challenges of using electric vehicles for promoting energy efficiency in Africa's clean energy transition. *Global Environmental Change, 84*, 102789. DOI: 10.1016/j.gloenvcha.2023.102789

Olabi, A. G., Wilberforce, T., Obaideen, K., Sayed, E. T., Shehata, N., Alami, A. H., & Abdelkareem, M. A. (2023). Micromobility: Progress, benefits, challenges, policy and regulations, energy sources and storage, and its role in achieving sustainable development goals. *International Journal of Thermofluids, 17*, 100292. DOI: 10.1016/j.ijft.2023.100292

Online, A. (2022). Egypt set to manufacture first electric vehicle in 2023, Sisi tells WYF. Retrieved from https://english.ahram.org.eg/NewsContent/1/1235/455019/Egypt/Urban--Transport/Egypt-set-to-manufacture-first-electric-vehicle-in.aspx. 11 January.

Othman, E., Cibilić, I., Poslončec-Petrić, V., & Saadallah, D. (2024). Investigating Noise Mapping in Cities to Associate Noise Levels with Sources of Noise Using Crowdsourcing Applications. *Urban Science (Basel, Switzerland)*, *8*(1), 13. DOI: 10.3390/urbansci8010013

Otto, A., Agatz, N., Campbell, J., Golden, B., & Pesch, E. (2018). Optimization approaches for civil applications of unmanned aerial vehicles (UAVs) or aerial drones: A survey. *Networks*, *72*(4), 411–458. DOI: 10.1002/net.21818

Page, S., & Ge, Y. (2009). Transportation and tourism: a symbiotic relationship? In *T. Jamal, & M. Robinson The SAGE handbook of tourism studies* (pp. 372–396). SAGE Publications Ltd., DOI: 10.4135/9780857021076.n21

Pan, S., Yu, W., Fulton, L. M., Jung, J., Choi, Y., & Gao, H. O. (2023). Impacts of the large-scale use of passenger electric vehicles on public health in 30 US. metropolitan areas. *Renewable & Sustainable Energy Reviews*, *173*, 113100. DOI: 10.1016/j.rser.2022.113100

Papadakis, D. M., Savvides, A., Michael, A., & Michopoulos, A. (2024). Advancing Sustainable Urban Mobility: Insights from Best Practices and Case Studies. *Fuel Communications*, *20*, 100125. DOI: 10.1016/j.jfueco.2024.100125

Papas, Th., Basbas, S., & Campisi, T. (2023). Urban mobility evolution and the 15-minute city model: From holistic to bottom-up approach. *Transportation Research Procedia*, *69*, 544–551. DOI: 10.1016/j.trpro.2023.02.206

Pardo-Bosch, F., Pujadas, P., Morton, C., & Cervera, C. (2021). Sustainable deployment of an electric vehicle public charging infrastructure network from a city business model perspective. *Sustainable Cities and Society*, *71*, 102957. DOI: 10.1016/j.scs.2021.102957

Patil, G., Pode, G., Diouf, B., & Pode, R. (2024). Sustainable Decarbonization of Road Transport: Policies, Current Status, and Challenges of Electric Vehicles. *Sustainability (Basel)*, *16*(18), 8058. DOI: 10.3390/su16188058

Pawlowska, Barbara & Suchanek, Michał. (2019). Transport as a Factor in the Achievement of the EU Goals to Combat Climate Changes and to Reduce Greenhouse Gases Emissions. Journal of Management and Financial Sciences. 79-96. DOI: 10.33119/JMFS.2018.33.5

Pelegov, D. V., & Pontes, J. (2018). Main drivers of battery industry changes: Electric vehicles—A market overview. *Batteries*, *4*(4), 65. DOI: 10.3390/batteries4040065

Pellegrino, F. (2021), "Transport and Tourism Relationship", Grasso, F. and Sergi, B.S. (Ed.) Tourism in the Mediterranean Sea, Emerald Publishing Limited, Bingley, pp. 241-256. DOI: 10.1108/978-1-80043-900-920211017

Perboli, G., Rosano, M., & Wei, Q. (2022). A simulation-optimization approach for the management of the on-demand parcel delivery in sharing economy. *IEEE Transactions on Intelligent Transportation Systems, 23*(8), 10570–10582. DOI: 10.1109/TITS.2021.3094851

Perera, C., Liu, C. H., & Jayawardena, S. (2017). The role of IoT in intelligent transport systems. *IEEE Internet of Things Journal, 4*(3), 857–868.

Pilati, F., Zennaro, I., Battini, D., & Persona, A. (2020). The sustainable parcel delivery (SPD) problem: Economic and environmental considerations for 3PLs. *IEEE Access : Practical Innovations, Open Solutions, 8*, 71880–71892. DOI: 10.1109/ACCESS.2020.2987380

Pojani, D., & Stead, D. (2017). *The urban transport crisis in emerging economies.* Springer. DOI: 10.1007/978-3-319-43851-1

Popescu, C. R. G., Yu, P., & Wei, Y. (1 C.E.). Achieving the Sustainable Development Goals Through Infrastructure Development. In *https://services.igi-global.com/resolvedoi/resolve.aspx?doi=10.4018/979-8-3693-0794-6*. IGI Global. https://www.igi-global.com/book/achieving-sustainable-development-goals-through/www.igi-global.com/book/achieving-sustainable-development-goals-through/324723

Pozo, R. F., González, A. B. R., Wilby, M. R., Vinagre Díaz, J. J., & Matesanz, M. V. (2022). Prediction of on-street parking level of service based on random undersampling decision trees. *IEEE Transactions on Intelligent Transportation Systems, 23*(7), 8327–8336. DOI: 10.1109/TITS.2021.3077985

Pupentsova, S., Demin, A., Kirilyuk, A., & Pupentsova, V. (2022). *The introduction of digital technologies to participatory design in the public spaces formation.* In Proceedings: Innovations in Digital Economy. Proceedings of the 3rd International Scientific Conference, SPBPU IDE 2021. Ser. 'Communications in computer and information science', 325-342. DOI: DOI: 10.1007/978-3-031-14985-6_23

Rajkumar, S., Sandhiya, S., & Manjusri, A. (2025). Optimized Traffic Routing System for Urban Congestion Management. *Journal of Ubiquitous Computing and Communication Technologies., 7*(1), 19–35. DOI: 10.36548/jucct.2025.1.002

Rani, S., & Jayapragash, R. (2024). Review on electric mobility: Trends, challenges and opportunities. *Results in Engineering, 102631*, 102631. Advance online publication. DOI: 10.1016/j.rineng.2024.102631

Rao, H., Zhang, D., Qin, G., Yue, L., & Sun, J. (2023). Investigating the intra-driver heterogeneity in car following behaviour based on large-scale naturalistic driving study. *Transportmetrica. B, Transport Dynamics*, *11*(1), 1363–1383. DOI: 10.1080/21680566.2023.2203848

Rastogi, R. (2011). Promotion of non-motorized modes as a sustainable transportation option: Policy and planning issues. *Current Science*, •••, 1340–1348.

Rebellion Research. (2022). Norway &. *Artificial Intelligence*, *23*(February). https://www.rebellionresearch.com/norway-artificial-intelligence

Ren, Y., Xie, R., Yu, F. R., Huang, T., & Liu, Y. (2022). Green intelligence networking for connected and autonomous vehicles in smart cities. *IEEE Transactions on Green Communications and Networking*, *6*(3), 1591–1603. DOI: 10.1109/TGCN.2022.3148293

Ribeiro, M. A., Gursoy, D., & Chi, O. H. (2022). Customer Acceptance of Autonomous Vehicles in Travel and Tourism. *Journal of Travel Research*, *61*(3), 620–636. DOI: 10.1177/0047287521993578

Riemens, R., Nast, C., Pelzer, P., & van den Hurk, M. (2021). An assessment framework for safeguarding public values on mobility platforms. *Urban Transformations*, *3*(1), 7. DOI: 10.1186/s42854-021-00023-3

Rietmann, N., & Lieven, T. (2019). How policy measures succeeded to promote electric mobility–Worldwide review and outlook. *Journal of Cleaner Production*, *206*, 66–75. DOI: 10.1016/j.jclepro.2018.09.121

Rizki, M.; Syahputri, J.; Belgiawan, P.F.; Irawan, M.Z. (2021). Electrifying Tourist Mobility in Bali, Indonesia: Setting the Target and Estimating the CO2 Reduction Based on Stated Choice Experiment. Sustainability 2021, 13, 11656. , 21 October. DOI: 10.3390/su132111656

Rodrigue, J. P. (2020). The Geography of Transport Systems. Retrieved from https://transportgeography.org/contents/applications/tourism-transport/

Rodrigue, J. P. (2020). *The geography of transport systems* (5th ed.). Routledge. DOI: 10.4324/9780429346323

Rodrigues, J.-P. (2024). *The Geography of Transport Systems* (6th ed.). Routledge., DOI: 10.4324/9781003343196

Rosenblat, A., & Stark, L. (2021). Algorithmic labor and information asymmetries: A case study of Uber's drivers. *International Journal of Communication*, *10*, 3758–3784.

Royce, R. (n.d.). The future is electric. Retrieved from https://www.rolls-royce.com/products-and-services/electrical.aspx

Ruggieri, R., Ruggeri, M., Vinci, G., & Poponi, S. (2021). Electric mobility in a smart city: European overview. *Energies*, *14*(2), 315. DOI: 10.3390/en14020315

Saberi, S., Kouhizadeh, M., & Sarkis, J. (2019). Blockchain technology in supply chain management. *International Journal of Production Research*, *57*(7), 2117–2135. DOI: 10.1080/00207543.2018.1533261

SAE International. (2021). *Taxonomy and Definitions for Terms Related to Driving Automation Systems for On-Road Motor Vehicles* (J3016_202104).

Sankar, M., Pachiyappan, S., Arjun, B. S., & Srivastava, A. (2024). Assessing the Impact of Smart City Technologies on Sustainable Urban Mobility in Developing Economies. In *Smart Cities and Circular Economy* (pp. 193–214). Emerald Publishing Limited., DOI: 10.1108/978-1-83797-957-820241018

Sankhwar, P. (2024). Suitability of Electric Vehicle Charging Infrastructure and Impact on Power Grid Due to Electrification of Roadway Transportation with Electric Vehicles. [IJSR]. *International Journal of Science and Research (Raipur, India)*, *13*(12), 374–383. DOI: 10.21275/SR241204033720

Santaella, J. J., Rodríguez-Bolívar, S., Puga-Pedregosa, L., González-Rico, A., Marín-González, M., & Gómez-Campos, F. M. (2022). High-luminance QD-LED device with digital and dynamic lighting functions for efficient automotive systems. *IEEE Photonics Journal*, *14*(2), 1–10. DOI: 10.1109/JPHOT.2022.3155650

Sawhney, N. (2023). Contestations in urban mobility: Rights, risks, and responsibilities for Urban AI. *AI & Society*, *38*(3), 1083–1098. DOI: 10.1007/s00146-022-01502-2

Schoemaker, P. J. H. (1995). Scenario planning: A tool for strategic thinking. *Sloan Management Review*, *36*(2), 25–40.

Schrank, D., Lomax, T., & Eisele, B. (2019). 2019 Urban mobility report. Texas Transportation Institute, [ONLINE]. Available: https://Static.Tti.Tamu.Edu/Tti.Tamu. Edu/Documents/Mobility-Report-2019.Pdf

Schwab, A., Thomas, A., Bennett, J., Robertson, E., & Cary, S. (2021). Electrification of Aircraft: Challenges, Barriers, and Potential Impacts. Golden, CO: National Renewable Energy Laboratory. NREL/TP-6A20-80220. https://www.nrel.gov/docs/fy22osti/80220.pdf

Scuttari, A., & Isetti, G. (2019). E-mobility and Sustainable Tourism Transport in Remote Areas. *Zeitschrift für Tourismuswissenschaft, 11*(2), 237–256. DOI: 10.1515/tw-2019-0014

Sengar, N., Kumari, I., Lee, J., & Har, D. (2024). Blockchain-based vision transformer network for safe driving perception. *IEEE Sensors Journal, 24*(20), 32907–32918. DOI: 10.1109/JSEN.2024.3454469

Seno, L. (2020). Autonomous Car Technology in PH Cars. Retrieved from https://www.philstar.com/the-freeman/cebu-business/2020/09/14/2042343/autonomous-car-technology-ph-cars, 14 September.

Shafiei, E., Davidsdottir, B., Fazeli, R., Leaver, J., Stefansson, H., & Asgeirsson, E. I. (2018). Macroeconomic effects of fiscal incentives to promote electric vehicles in Iceland: Implications for government and consumer costs. *Energy Policy, 114*, 431–443. DOI: 10.1016/j.enpol.2017.12.034

Shah, K. J., Pan, S. Y., Lee, I., Kim, H., You, Z., Zheng, J. M., & Chiang, P. C. (2021). Green transportation for sustainability: Review of current barriers, strategies, and innovative technologies. *Journal of Cleaner Production, 326*, 129392. DOI: 10.1016/j.jclepro.2021.129392

Sharma, R., & Ghosh, D. (2021). Smart urban transport: Predictive analytics for transit management. *International Journal of Transportation Science and Technology, 10*(1), 1–11.

Shaw, C., Hales, J., & Li, B. (2020). Gender and urban mobility: Perspectives from low-income communities. *Journal of Transport Geography, 87*, 102813.

Shaygan, M., Meese, C., Li, W., Zhao, X. G., & Nejad, M. (2022). Traffic prediction using artificial intelligence: Review of recent advances and emerging opportunities. *Transportation Research Part C, Emerging Technologies, 145*, 103921. DOI: 10.1016/j.trc.2022.103921

Shevchuk, A. V., Chizhova, D. A., Demina, D. E., & Chugunkova, V. A. (2021). Taxi driver operation under the algorithmic control. *Monitoring, 3*, 356–381.

Silva, Ó., Cordera, R., González-González, E., & Nogués, S. (2022). Environmental impacts of autonomous vehicles: A review of the scientific literature. *The Science of the Total Environment, 830*, 154615. https://doi.org/https://doi.org/10.1016/j.scitotenv.2022.154615. DOI: 10.1016/j.scitotenv.2022.154615 PMID: 35307440

Singh, A. R., Vishnuram, P., Alagarsamy, S., Bajaj, M., Blazek, V., Damaj, I., Rathore, R. S., Al-Wesabi, F. N., & Othman, K. M. (2024). Electric vehicle charging technologies, infrastructure expansion, grid integration strategies, and their role in promoting sustainable e-mobility. *Alexandria Engineering Journal*, *105*, 300–330. DOI: 10.1016/j.aej.2024.06.093

SkillsFuture SG. (2021). *About SkillsFuture*.

Sluiter, C. (2021). The new EU Sustainable and Smart Mobility Strategy: A local and regional perspective https://www.cidob.org/sites/default/files/2024-07/25-30_CASPAR%20SLUITER_ANG.pdf

Smith, A., & Anderson, J. (2021). *AI and the future of transportation jobs*. Pew Research Center.

Smith, J., & Brown, K. (2020). Predictive maintenance in transportation networks. *Engineering & Maintenance Review*, *5*(2), 88–101.

Snarveier. (2016). Norwegian EV policy. Retrieved from https://elbil.no/english/norwegian-ev-policy/, 13 May.

Song, Z., Liu, L., Jia, F., Luo, Y., Jia, C., Zhang, G., Yang, L., & Wang, L. (2024). Robustness-aware 3D object detection in autonomous driving: A review and outlook. *IEEE Transactions on Intelligent Transportation Systems*, *25*(11), 15407–15436. DOI: 10.1109/TITS.2024.3439557

Sony, M., & Naik, S. (2020). Industry 4.0 integration with socio-technical systems theory: A systematic review and proposed theoretical mode. *Technology in Society*, *61*, 10124. DOI: 10.1016/j.techsoc.2020.101248

Subrahmanyam, S. (2025). Data Governance and Privacy in Smart Cities. In Sharma, A., Mansotra, V., & Singh, P. (Eds.), *Citizen-Centric Artificial Intelligence for Smart Cities* (pp. 433–456). IGI Global Scientific Publishing. DOI: 10.4018/979-8-3693-7832-8.ch017

Subrahmanyam, S. (2025). Employee Skillsets in the Digital Age. In Özbek, M., & Christiansen, B. (Eds.), *Organizational Sociology in the Digital Age* (pp. 369–388). IGI Global Scientific Publishing. DOI: 10.4018/979-8-3693-7398-9.ch018

Subrahmanyam, S. (2025). Role of Artificial Intelligence in Re-Inventing Human Resource Management (HRM) 2.0. In Kapoor, T., Singh, A., Sahay, K., Kaur, H., & Verma, R. (Eds.), *Sustainable Management Practices for Employee Retention and Recruitment* (pp. 243–264). IGI Global Scientific Publishing. DOI: 10.4018/979-8-3693-8679-8.ch012

Subrahmanyam, S., Azoury, N., & Sarkis, N. (2024). Optimising Energy Efficiency through Effective Supply Chain Management: A Comparative Study. *ACTA INNOVATIONS*, *51*(1), 46–51. DOI: 10.62441/ActaInnovations.51.7

Subrahmanyam, S., Khalife, D., & El-Chaarani, H. (2024). Towards Sustainable Future: Exploring Renewable Energy Solutions and Environmental Impacts. *ACTA INNOVATIONS*, *51*(1), 15–24. DOI: 10.62441/ActaInnovations.51.3

Sultana, S., Salon, D., & Kuby, M. (2019). Transportation sustainability in the urban context: A comprehensive review. *Urban Geography*, *40*(3), 279–308. DOI: 10.1080/02723638.2017.1395635

Sumitkumar, R., & Al-Sumaiti, A. S. (2024). Shared autonomous electric vehicle: Towards social economy of energy and mobility from power-transportation nexus perspective. *Renewable & Sustainable Energy Reviews*, *197*, 114381. DOI: 10.1016/j.rser.2024.114381

Suvvari, S., & Saxena, V. D. (2023). Stakeholder Management in Projects: Strategies for Effective Communication. *Innovative Research Thoughts.*, *9*(5), 188–201. DOI: 10.36676/irt.v9.i5.1479

Szpilko, D., Budna, K., & Drmeyan, H. Remiszewska, (2023). Sustainable and smart mobility—Research directions: A systematic literature review. Ekon. Sr, 3, 31–61. ttps://DOI: <ALIGNMENT.qj></ALIGNMENT>10.34659/eis.2023.86.3.584

Tang, J., Wang, Y., Liu, W., Luo, X., & Zhou, Z. (2021). SCP-Tree: Finding multiple nearest parking spots with minimal group travel cost. *IEEE Access : Practical Innovations, Open Solutions*, *9*, 158946–158960. DOI: 10.1109/ACCESS.2021.3131229

Taylor, B., Anderson, D., & Carter, P. (2021). Autonomous vehicles in smart cities. *Urban Mobility Innovations*, *14*(1), 29–47.

Te Manatu Wakak Ministry of Transport. (July 2021). Electric Vehicles Programme. Retrieved from https://www.transport.govt.nz/area-of-interest/environment-and-climate-change/electric-vehicles-programme/#:~:text=Programme%20announced,by%20the%20end%20of%202021

The World Bank. Urban Development. 2017. Available online: https://www.worldbank.org/en/topic/urbandevelopment/overview#:~:text=Today%2C%20more%20than%20half%20of,people%20will%20live%20in%20cities. (accessed on 5 January 2024)

Trencher, G., & Wesseling, J. (2022). Roadblocks to fuel-cell electric vehicle diffusion: Evidence from Germany, Japan and California. *Transportation Research Part D, Transport and Environment*, *112*, 103458. DOI: 10.1016/j.trd.2022.103458

Tu, J. C., & Yang, C. (2019). Key factors influencing consumers' purchase of electric vehicles. *Sustainability (Basel), 11*(14), 3863. DOI: 10.3390/su11143863

Tuli, F. M., Mitra, S., & Crews, M. B. (2021). Factors influencing the usage of shared e-scooters in Chicago. *Transportation Research Part A, Policy and Practice, 154*, 164–185. DOI: 10.1016/j.tra.2021.10.008

Tundys, B., & Wiśniewski, T. (2023). Smart Mobility for Smart Cities—Electromobility Solution Analysis and Development Directions. *Energies, 16*(4), 1958. DOI: 10.3390/en16041958

U.S. Department of Energy. (n.d.). Electricity Laws and Incentives in Federal. Retrieved from https://afdc.energy.gov/fuels/laws/ELEC?state=US

U.S. Department of Transportation. (2019). U.S. DOT Artificial Intelligence Activities. Retrieved from https://www.transportation.gov/AI, 23 September.

Udacity. (2021). *Mercedes-Benz and Udacity: Collaborating on autonomous systems education.*

Umali, T. (2019). The Philippines to use autonomous vehicles during SEA Games. Retrieved from https://opengovasia.com/the-philippines-to-use-autonomous-vehicles-during-sea-games/, 17 July.

United Nations Economic Commission for Europe (UNECE). (2020). *Framework for automated/autonomous vehicles.*

United Nations Economic Commission for Europe. (n.d.). Climate Change and Sustainable Transport. Retrieved from https://unece.org/climate-change-and-sustainable-transport

United Nations New York. (2020). World Economic Situation and Prospects. Retrieved from https://www.un.org/development/desa/dpad/wp-content/uploads/sites/45/WESP2020_Annex.pdf

United Nations. (1992). United Nations Framework Convention on Climate Change. Retrieved from https://unfccc.int/files/essential_background/background_publications_htmlpdf/application/pdf/conveng.pdf

United Nations. (n.d.). Climate Action. Retrieved from https://www.un.org/en/climatechange/climate-solutions/cities-pollutio

United States Environmental Protection Agency. (2019). Sources of Greenhouse Gas Emissions. Retrieved from https://www.epa.gov/ghgemissions/sources-greenhouse-gas-emissions

United States Environmental Protection Agency. (n.d.). Strategies for Climate Change Adaptation. Retrieved from https://www.epa.gov/arc-x/strategies-climate-change-adaptation

University of Surrey. (2018, November 7). Autonomous vehicles could shape the future of urban tourism. *ScienceDaily*. Retrieved from www.sciencedaily.com/releases/2018/11/181107111845.htm, 29 December.

Un-Noor, F., Padmanaban, S., Mihet-Popa, L., Mollah, M. N., & Hossain, E. (2017). A comprehensive study of key electric vehicle (EV) components, technologies, challenges, impacts, and future direction of development. *Energies*, *10*(8), 1217. DOI: 10.3390/en10081217

Uribe-Chavert, P., Posadas-Yagüe, J.-L., & Poza-Luján, J.-L. (2025). Evaluating Traffic Control Parameters: From Efficiency to Sustainable Development. *Smart Cities.*, *8*(2), 57. DOI: 10.3390/smartcities8020057

USDOT. (2022). *Transportation workforce development and training programs*. U.S. Department of Transportation.

Van Truong, N., & Shimizu, T. (2017). The effect of transportation on tourism promotion: Literature review on application of the Computable General Equilibrium (CGE) *Model. Transportation Research Procedia*, *25*, 3096–3115. DOI: 10.1016/j.trpro.2017.05.336

Varghese, A. M., Menon, N., & Ermagun, A. (2024). Equitable distribution of electric vehicle charging infrastructure: A systematic review. *Renewable & Sustainable Energy Reviews*, *206*, 114825. DOI: 10.1016/j.rser.2024.114825

Vashkevich, N., Barykin, S. E., Krasnov, A., Kurochkina, A. A., Alekseeva, N., Krasnova, D. S., Kankhva, V. S., Khovalova, T. V., Mikhov, O., Buniak, V., Dzhamaludinova, M. Y., Dedyukhina, N., & Ali, M. (2024). E-government based on sustainable digitalization: The case of the Chinese social sustainability marketing technologies paradigm. *Journal of Infrastructure. Policy and Development*, *8*(3), 3. Advance online publication. DOI: 10.24294/jipd.v8i3.3113

Verbong, G. P. J., & Geels, F. W. (2010). Exploring sustainability transitions in the electricity sector with socio-technical pathways. *Technological Forecasting and Social Change*, *77*(8), 1214–1221. DOI: 10.1016/j.techfore.2010.04.008

Vijayaraghavan, V., & Rian Leevinson, J. (2020). Intelligent Traffic Management Systems for Next Generation IoV in Smart City Scenario. In Mahmood, Z. (Ed.), *Connected Vehicles in the Internet of Things: Concepts, Technologies and Frameworks for the IoV* (pp. 123–141). Springer International Publishing., DOI: 10.1007/978-3-030-36167-9_6

Visaria, A. A., Jensen, A. F., Thorhauge, M., & Mabit, S. E. (2022). User preferences for EV charging, pricing schemes, and charging infrastructure. *Transportation Research Part A, Policy and Practice*, *165*, 120–143. DOI: 10.1016/j.tra.2022.08.013

Vorwerg, F., Torkayesh, A., & Venghaus, S. (2025). From fossil fuels to alternative fuels: Strategy development for a sustainable transport sector in Germany. *Energy, Sustainability and Society*, *15*(1), 4. Advance online publication. DOI: 10.1186/s13705-024-00498-5

Vuković, D., Vlasova, N. Y., & Ergunova, O. T. (2020). Institutions in the Context of the Regional Development of Russia. In Gabdrakhmanov, N. K., & Safiullin, L. N. (Eds.), *Regional Economic Development in Russia* (pp. 123–132). Springer International Publishing., DOI: 10.1007/978-3-030-39859-0_11

Wang, L. & Shang, A. & Feng, Y. & Fu, H. & Gao, Z. (2025). Research on Regional Traffic Simulation Analysis and Traffic Capacity Improvement Technology Under the Background of Urban Renewal. E3S Web of Conferences. 617. DOI: 10.1051/e3sconf/202561702001

Wang, H. M., Avedisov, S. S., Altintas, O., & Orosz, G. (2024). Intent sharing in cooperative maneuvering: Theory and experimental evaluation. *IEEE Transactions on Intelligent Transportation Systems*, *25*(9), 12450–12463. DOI: 10.1109/TITS.2024.3379994

Wang, J., Xu, B., & Zhou, L. (2023). Resilience analysis of multi-modal networks using graph-theoretic approaches. *Transportation Research Part C, Emerging Technologies*, *138*, 1035–1049.

Wang, J., Yu, H., & Zhao, M. (2023). Quantum computing in transportation network optimization: Current trends and future directions. *Transportation Research Part C, Emerging Technologies*, *145*, 103921.

Wang, M., Zhou, L., & Feng, X. (2023). IoT-enabled smart logistics for multi-modal transport. *Future Transportation Journal*, *13*(2), 75–90.

Wang, S., Wang, Z., Jiang, R., Yan, R., & Du, L. (2022). Trajectory jerking suppression for mixed traffic flow at a signalized intersection: A trajectory prediction based deep reinforcement learning method. *IEEE Transactions on Intelligent Transportation Systems, 23*(10), 18989–19000. DOI: 10.1109/TITS.2022.3152550

Wang, Y., Zhu, X., Li, L., & Wu, B. (2013). Reasons and countermeasures of traffic congestion under urban land redevelopment. *Procedia: Social and Behavioral Sciences, 96*, 2164–2172. DOI: 10.1016/j.sbspro.2013.08.244

Weiss, M., Cloos, M., & Kasten, P. (2020). *The impact of electrification on the German automotive workforce*. Fraunhofer Institute.

Wes, M. (19 April 2022). Egypt: Acting Against Climate Change for A Healthier, More Prosperous Future. https://www.worldbank.org/en/news/opinion/2022/04/19/-egypt-acting-against-climate-change-for-a-healthier-more-prosperous-future

Wolniak, R. (2023). Smart mobility in a smart city concept. Scientific Papers of Silesian University of Technology. *Organization and Management Series., 2023*(170), 679–692. Advance online publication. DOI: 10.29119/1641-3466.2023.170.41

World Bank. (2022). *Transforming transportation: Smart mobility for sustainable cities*.

World Economic Forum. (2020). *The future of jobs report 2020*.

World Tourism Organization and International Transport Forum. (2019). Transport-related CO2 Emissions of the Tourism Sector – Modelling Results. UNWTO. Madrid, DOI: DOI: 10.18111/9789284416660

World Tourism Organization and United Nations Development Programme. (2017). *Tourism and the Sustainable Development Goals – Journey to 2030*. UNWTO.

World Tourism Organization. (2021). *International Tourism Highlights* (2020 Edition). UNWTO., DOI: 10.18111/9789284422456

World Tourism Organization. (2021). UNWTO Secretary-General: 'The Climate Emergency Is a Bigger Threat Than Covid' https://www.unwto.org/news/unwto-secretary-general-the-climate-emergency-is-a-bigger-threat-than-covid, 3 November.

World Tourism Organization. (2022). Tourism Enjoys Strong Start to 2022 while Facing New Uncertainties. Retrieved from https://www.unwto.org/news/tourism-enjoys-strong-start-to-2022-while-facing-new-uncertainties, 25 March.

World Tourism Organization. (n.d). Tourism – an economic and social phenomenon. retrieved from https://www.unwto.org/why-tourism

World Tourism Organization. (n.d). Transforming Tourism for Climate Action. Retrieved from https://www.unwto.org/sustainable-development/climate-action#:~:text=The%20tourism%20sector%20is%20highly,the%20resilience%20of%20the%20sector

Wurman, P. R., D'Andrea, R., & Mountz, M. (2008). Coordinating hundreds of cooperative, autonomous vehicles in warehouses. *AI Magazine*, *29*(1), 9–20.

Wüstenhagen, S., & Kirschstein, T. (2024). Substitution of Conventional Vehicles in Municipal Mobility. *Sustainability (Basel)*, *16*(14), 6054. DOI: 10.3390/su16146054

Wu, X., Liu, Y., & Zhu, X. (2020). Modeling and analysis of multimodal transportation networks. *Transportation Research Part B: Methodological*, *135*, 1–18.

Yan, X., Allison, C. K., Fleming, J. M., Stanton, N. A., & Lot, R. (2021). The benefit of assisted and unassisted eco-driving for electrified powertrains. *IEEE Transactions on Human-Machine Systems*, *51*(4), 403–407. DOI: 10.1109/THMS.2021.3086057

Yuan, Y., Liu, P., & Yang, X. (2019). Graph-based models for multi-modal transportation planning: A survey. *Transportation Research Part B: Methodological*, *129*, 26–50.

Yu, H., Jiang, R., He, Z., Zheng, Z., Li, L., Liu, R., & Chen, X. (2021). Automated vehicle-involved traffic flow studies: A survey of assumptions, models, speculations, and perspectives. *Transportation Research Part C, Emerging Technologies*, *127*, 103101. https://doi.org/https://doi.org/10.1016/j.trc.2021.103101. DOI: 10.1016/j.trc.2021.103101

Yu, J. J., Tang, C. S., Li, M. K., & Shen, Z. J. M. (2022). Coordinating installation of electric vehicle charging stations between governments and automakers. *Production and Operations Management*, *31*(2), 681–696. DOI: 10.1111/poms.13564

Zang, G., Azouigui, S., Saudrais, S., Hébert, M., & Gonçalves, W. (2022). Evaluating the understandability of light patterns and pictograms for autonomous vehicle-to-pedestrian communication functions. *IEEE Transactions on Intelligent Transportation Systems*, *23*(10), 18668–18680. DOI: 10.1109/TITS.2022.3165881

Zemmouchi-Ghomari, L. (2025). Artificial intelligence in intelligent transportation systems. Journal of Intelligent Manufacturing and Special Equipment. https://doi.org/DOI: 10.1108/JIMSE-11-2024-0035

Zentani, A., Almaktoof, A., & Kahn, M. T. (2024). A Comprehensive Review of Developments in Electric Vehicles Fast Charging Technology. *Applied Sciences (Basel, Switzerland)*, *14*(11), 4728. DOI: 10.3390/app14114728

Zhang, R. (2022). Decarbonizing Japan's Transportation Sector. Retrieved from https://www.nbr.org/publication/decarbonizing-japans-transportation-sector-toward-net-zero-emissions/, 12 April.

Zhang, R. and Fujimori, S. (2020). The role of transport electrification in global climate change mitigation scenarios. Environ. Res. Lett. 15 034019, 19 February.

Zhang, Y., Chen, Y., Li, X., Wang, Y., Wu, X., & Gao, Y. (2020). Journal of Cleaner Production. *A Review of Battery Recycling and Its Effect on Environment and Economy, 253*, 119886.

Zhang, C., Yan, J., & You, F. (2023). Critical metal requirement for clean energy transition: A quantitative review on the case of transportation electrification. *Advances in Applied Energy*, *9*, 100116. DOI: 10.1016/j.adapen.2022.100116

Zhang, J., Wang, X., & Chen, Z. (2022). Big data-driven traffic flow prediction using deep learning techniques. *IEEE Transactions on Intelligent Transportation Systems*, *23*(5), 3045–3057.

Zhang, S., Li, J., Shi, L., Ding, M., Nguyen, D. C., Tan, W., Weng, J., & Han, Z. (2023). Federated Learning in Intelligent Transportation Systems: Recent Applications and Open Problems. arXiv:2309.11039. https://arxiv.org/abs/2309.11039

Zhang, W., Barykin, S., Kirillova, T., Kapustina, I., Lukashevich, N., & Zaytsev, A. (2023). An Innovative Resource Management Framework using Logit-Boosted Machine Learning Algorithms for Vehicular Ad Hoc Networks (VANETs). *Engineering and Science*. Advance online publication. DOI: 10.30919/es980

Zhang, X., Zhang, Y., & Chen, A. (2020). A resilience-based approach for multimodal transportation network robustness evaluation. *Transportation Research Part E, Logistics and Transportation Review*, *136*, 101887.

Zhang, Y., Liu, H., & Wang, T. (2021). Multi-modal transportation networks: An overview and future directions. *Journal of Translational Science*, *17*(1), 1–20.

Zhang, Y., Liu, Z., & Chen, J. (2021). The impact of AI on multi-modal transportation planning. *Journal of Transportation Engineering*, *147*(4), 401–419.

Zhang, Y., Wang, J., & Zhao, J. (2021). Smart mobility: Exploring the impact of intelligent transportation systems on urban transport sustainability. *Journal of Transport Geography*, *94*, 103134.

Zhao, P., Wang, F., & Sun, Y. (2022). Graph neural networks for dynamic transport network modeling. *Neural Computing & Applications*, *34*(9), 14234–14249.

Zhao, Q., He, Y., & Chen, L. (2022). Smart grid integration for sustainable transport systems. *Renewable Energy & Mobility*, *8*(3), 120–145.

Zhao, X., Li, X., Jiao, D., Mao, Y., Sun, J., & Liu, G. (2024). Policy incentives and electric vehicle adoption in China: From a perspective of policy mixes. *Transportation Research Part A, Policy and Practice*, *190*, 104235. DOI: 10.1016/j.tra.2024.104235

Zhao, Y., Gao, H., Zhang, L., & Wang, Y. (2019). Intelligent transportation systems: When AI meets transportation. *Transportation Research Part C, Emerging Technologies*, *100*, 1–4. DOI: 10.1016/j.trc.2019.01.013

Zheng, Z., Wang, Z., Zhu, L., & Jiang, H. (2020). Determinants of the congestion caused by a traffic accident in urban road networks. *Accident; Analysis and Prevention*, *136*, 105327. DOI: 10.1016/j.aap.2019.105327 PMID: 31874332

Zhu, W., & Zhu, C. (2023). A prediction-based route guidance method toward intelligent and green transportation system. *IEEE Internet of Things Journal*, *10*(14), 12763–12776. DOI: 10.1109/JIOT.2023.3255200

About the Contributors

Olga Ergunova is a renowned faculty in Business Management. Presently working as Associate Professor in Associate Professor, Higher School of Production, she is also a member of various jury at the international level and expert of the Russian Academy of Sciences.

Dinesh Elango is a highly experienced academic professional with over 12 years in the field, Dr. Dinesh Elango is a respected Associate Professor of Business Analytics and Business Management. With a PhD in Management Sciences from Anna University, Chennai, Dr. Dinesh has been recognized as an Accredited Management Teacher (AMT) by the All India Management Association.

Pawan Kumar is a Professor in Business Managment having 20 years of experience in teaching, Industry and research. He is in the editorial board member of inderscience journals and handling edited books from IGI, Emerald, Taylor Francis etc.

Preeti Mehra is Professor in Marketing with an expertise in public speaking, academic instruction, and research management. Offers history of delivering marketing advancements and significant university curricula development. She is devoted to creating future field leaders through education and mentoring.

Rajesh Verma is a Professor & Sr. Dean at Mittal School of Business (NIRF Ranked), Lovely Professional University (LPU), where he leads the academic and administrative functions of the business school. He has a PhD in Marketing from Himachal Pradesh University, an MBA from Utkal University, and an Executive Programme in Data Driven Product Management from IIM Lucknow. He is also a Wiley Certified Design Thinking Practitioner

***

Santosh Adhikari is pursuing a Master's degree in Civil Engineering with a focus on Transportation Engineering. He has a strong interest in traffic safety, roadway design, and data-driven analysis. His research explores crash trends, vulnerable road user safety, and roundabout operations. Santosh combines engineering skills with GIS, statistics, and real-world problem solving.

Rashid Ating's research area is about financial ageing and retirement, financial literacy, healthcare, tourism, and data visualization. His research outputs were published both in English and his national language. He likes to jog, bake cookies, fly a drone, play badminton, and become a solo traveller

Arbia Hlali is a doctor of economics science from the Faculty of Economics and Management of Sfax, University of Sfax, Tunisia. She holds a master of transport and logistics sciences from the higher Institute of industrial management of Sfax, Tunisia. Currently, she is an Assistant Professor at Taibah University, Saudi Arabia. She has international experience in teaching and researching and earlier worked at University of Puerto Rico at Rio Piedras, San Juan, Puerto Rico, and prior to that at different Universities in Tunisia. Her research interests include Logistics, Supply Chain, Transport Economics, Management, Maritime Transport, Transport Management, Container Shipping, Transportation Science, Containers port, Port Management, Business Model Innovation, Advanced Econometrics, Standard Statistical Analysis, Operation Management, strategic Management, Efficiency, Sustainability, among others. She is a prolific researcher and has several publications in renowned journals and books. She work also as editor and reviewer in several international journals.

Maxim Izmaylov is an Associate Professor. Research interests include: management theory, production management, sustainable development. Author and co-author of more than 150 publications, participant of international conferences. Research supervisor of more than 20 bachelor and masters' qualification works.

Viliyan Krastev, Doctor of Social Sciences (DSc), Candidate of Geographical Sciences (PhD), Associate Professor. Research interests include: geography of tourist flows.

April Kim is an assistant professor at the College of International Tourism and Hospitality Management of the University of Perpetual Help System DALTA Las Piñas, Philippines. She is currently taking her Ph.D. in International Tourism Management at Lyceum of the Philippines University Manila. She is also an alumna of the Philippine Women's University Manila and the University of the Philippines - Diliman. Her research interests are accessible tourism, responsible tourism, and regenerative tourism.

Maria Livintsova has a PhD in Economics, Associate Professor. Research interests include: management theory, production management, lean production, sustainable development. Author and co-author of more than 100 publications, member of the organizing committee and participant of international conferences. Research supervisor of more than 40 bachelor and masters' qualification works.

Habib Noorbhai obtained his Ph.D. in Exercise Science from the University of Cape Town and currently serves as a Professor (Health & Sports Science) and Director of the Biomedical Engineering and Healthcare Technology (BEAHT) Research Centre at the University of Johannesburg. He is involved with inter-disciplinary work in the healthcare technology and innovation spaces and has a particular interest on how health sciences students can advance their learning and skills using methods and tools through extended realities (XR). In 2017, he won the Mr South Africa title, and used this platform to spearhead change within health, sport and education. He was later inducted as a Golden Key International Honorary Member at UCT in 2018 for his contributions to academia, leadership and service. In 2021, he was awarded the Distinguished Vice-Chancellor's Award for Teacher Excellence at UJ. Prof. Noorbhai holds an NRF Y-rating, and continues to conduct research within the cricket sciences. He has published over 50 papers in local and international journals (research articles, conference proceedings and book chapters)

Yury Nurulin graduated with honors from Leningrad Polytechnic University in 1978. From 1982 to the present, he has been working at Peter the Great St. Petersburg Polytechnic University. In 1986, he defended his PhD thesis in the specialty "Automation of technological processes and production. In 1999, he received the academic degree of Doctor of Technical Sciences. Scientific supervisor of the master's program "Energy Management". Author of more than 90 scientific papers, including 7 books. In 2012-2022, he was the head of 9 international projects of the European programs BSR INTERREG and CBC. Research interests - innovation management, energy management, socio-technical systems

Renu P Pathak is an Associate Professor and Head, Department of Mathematics, SoS, Sandip University, Nasik, India. She got her Masters and PhD. in Mathematics from Barkatullah university, Bhopal, M.P. Her research interest is in Fixed point theory. This includes knowledge real analysis, topology and its applications in applied mathematics. Her research has been disseminated in several internationally reputed journals and conferences. Eight Ph.D. Scholars are taken a guidance from her in which four scholars completed their RAC. Also three students awarded Ph.D under her guidance and two are submitted their thesis. In her work we published two patents and various research papers in journals also presented papers in various national and international conferences

Animesh Kumar Sharma is a Research Scholar at Mittal School of Business, Lovely Professional University, Phagwara, Punjab, India and working as Manager - Marketing and Corporate Communication with Vatika Business Centres Private Limited (A Vatika Group Company), Gurugram, India. His research interest areas include digital marketing, social media marketing, search engine marketing, artificial intelligence, machine learning, data analytics and the applications of technology in business.

Rahul Sharma is a highly accomplished professor of Marketing with over 14 years of experience in academia. He has a Ph.D. in Marketing and has published over 15 articles in high-quality journals in the field. Dr. Sharma's research interests include consumer behavior, business analytics, and digital marketing. In addition to his research, Dr. Rahul is also a highly sought-after resource person in various faculty development programs.

Galina Silkina, Doctor of Economics, Professor, Professor. Research interests include: Innovative processes in the digital economy. Mathematical models of socio-economic processes. Logistics. Research supervisor of more than 10 post-graduate students, and 50 bachelor and masters' qualification works. Participant of domestic and international conferences.

Inga Skvortsova graduated from the St. Petersburg State Technical University with a degree in Economics and Management in the Fuel and Energy Sector. In 2004, she defended her dissertation for the degree of candidate of economic sciences in the specialty "Economics and management of the national economy: economics, organization and management of enterprises, industries, complexes (industry)". From 2004 to the present, she has been working at Peter the Great St. Petersburg Polytechnic University, associate professor. In 2024, she received the degree of Doctor of Technical Sciences. Author of more than 50 scientific papers, including 3 books. Research interests: methods and tools of innovation management, energy management, economics of enterprises, socio-technical systems

Satya Subrahmanyam is a distinguished academic leader, researcher, and institutional strategist with over 27 years of experience in higher education, leadership, and research management. He holds two PhDs and a Postdoctoral qualification from the Polytechnic Institute of Leiria, Portugal. Currently serving as Professor and Dean (Research) at Holy Spirit University of Kaslik (USEK), Lebanon, he spearheads research initiatives that enhance academic excellence and institutional reputation. He played a pivotal role in the development of the Catholic University in Erbil, Kurdistan, where he established new departments, led research programs, and collaborated on international educational projects. As a founding member and Managing Partner of Vignan Institute of Technology and Management, India, he contributed to its transformation into a premier institution. With expertise in AI-driven business strategies, HR analytics, and leadership in technology adoption, he has published extensively, guided PhD research, and led training programs in corporate and academic settings.

V. Padmavathi, M.E., Ph.D., is an Associate Professor in the Department of Information Technology at A.V.C. College of Engineering, Mannampandal, Mayiladuthurai. With over 16 years of teaching experience, she has contributed to research in areas such as Internet of Things, Computer Networks, and Cyber Threat Intelligence. She also served as a resource person for various workshops and programs. Her dedication to education and technology underscores her valuable contributions to the field. Her academic contributions include research papers, book chapters, book publication, patents, and conference presentations. She has conducted ICSSR Sponsored National Seminars during 2021-22, 2022-23 and 2023-24, DRDO Sponsored National Seminars during 2023-24. Her commitment to education extends beyond the classroom, as she actively contributes to workshops and programs. Author's full name: Padmavathi V Department: Information Technology Affiliation details: A.V.C.College of Engineering Mail id:vvpadhu@gmail.com

Tatyana Zorina has many years of experience in participating in national and international projects for the development of the energy complex. Member of the International Association for Energy Economics (IAEE). Under the leadership of prof. Zorina, the scientific school on Sustainable Energy Development of the Republic of Belarus successfully operates.

Index

A

B

C

D

E

G

I

U

V

W

www.ingramcontent.com/pod-product-compliance
Lightning Source LLC
LaVergne TN
LVHW061542290926
849162LV00004B/877
9798337308821